HELP YOUR KIDS WITH
maths

HELP YOUR KIDS WITH

m$\sqrt[a]{t}$hS

A UNIQUE STEP-BY-STEP VISUAL GUIDE

LONDON, NEW YORK, MELBOURNE,
MUNICH, AND DELHI

Project Art Editor
Mark Lloyd

Project Editor
Nathan Joyce

Designers
Nicola Erdpresser, Riccie Janus,
Maxine Pedliham, Silke Spingies,
Rebecca Tennant

Editors
Nicola Deschamps, Martha Evatt,
Lizzie Munsey, Martyn Page, Laura Palosuo,
Peter Preston, Miezan van Zyl

Design Assistants
Thomas Howey, Fiona Macdonald

Indexer
Jane Parker

Production Editor
Luca Frassinetti

Production
Erica Pepe

Jacket Designer
Duncan Turner

Managing Editor
Sarah Larter

Managing Art Editor
Michelle Baxter

Publishing Manager
Liz Wheeler

Art Director
Phil Ormerod

Reference Publisher
Jonathan Metcalf

First published in Great Britain in 2010 by
Dorling Kindersley Limited
80 Strand, London WC2R 0RL

A Penguin Company

Copyright © 2010 Dorling Kindersley Limited

8 10 9 7
019 – TD339 – Jul/2010

A CIP catalogue record for this book is
available from the British Library.

ISBN: 978 1 4053 2246 1

Printed in China.

See our complete catalogue at
www.dk.com

CAROL VORDERMAN M.A.(Cantab), MBE is one of Britain's best loved TV presenters and is renowned for her skills in mathematics. She has hosted numerous shows from light entertainment with **Carol Vorderman's Better Homes** and **The Pride of Britain Awards**, to scientific programmes such as **Tomorrow's World**, on the BBC, ITV and Channel 4. Whether co-hosting Channel 4's **Countdown** for 26 years, becoming the second best-selling female non-fiction author of the noughties decade in the UK, advising Rt Hon David Cameron on the future of potential mathematics education in the UK, Carol has a passion and devotion to explaining mathematics in an exciting and easily understandable way. In 2010 she launched her own online maths school **www.themathsfactor.com** where she teaches parents and children how they can become the very best they can be in the art of arithmetic.

BARRY LEWIS (Consultant Editor, Numbers, Geometry, Trigonometry, Algebra) read mathematics at university and graduated with a first class honours degree. He spent many years in publishing, as an author and as an editor, where he developed a passion for mathematical books that presented this often difficult subject in accessible, appealing, and visual ways. Among these is **Diversions in Modern Mathematics**, which subsequently appeared in Spanish as **Matemáticas modernas. Aspectos recreativos.**

He was invited by the British Government to run the major initiative **Maths Year 2000**, a celebration of mathematical achievement with the aim of making the subject more popular and less feared. In 2001 Barry became the President of The Mathematical Association, and for his achievements in popularizing mathematics he was elected a Fellow of the Institute of Mathematics and its Applications. He is currently the Chair of Council of The Mathematical Association and regularly publishes articles and books dealing with both research topics and ways of engaging people in this critical subject.

ANDREW JEFFREY (Probability) is a maths consultant, well known for his passion and enthusiasm for the teaching and learning of mathematics. A teacher and inspector for over 20 years, Andrew now spends his time training, coaching, and supporting teachers and delivering lectures for various organizations throughout Europe. Andrew's previous books include **Magic Maths for Kids**, **Top 20 Maths Displays**, **100 Top Tips for Top Maths Teachers**, and **Be a Wizard With Numbers**. Andrew is also better known to many schools as the Mathemagician, delivering his "Magic of Maths" shows to young and old! **www.andrewjeffrey.co.uk**.

MARCUS WEEKS (Statistics) is the author of many books and has contributed to several encyclopedias, including DK's **Science: The Definitive Visual Guide** and **Children's Illustrated Encyclopedia**.

Contents

FOREWORD by Carol Vorderman   8
INTRODUCTION by Barry Lewis   10

1 NUMBERS

Introducing numbers	14
Addition	16
Subtraction	17
Multiplication	18
Division	22
Prime numbers	26
Units of measurement	28
Positive and negative numbers	30
Powers and roots	32
Standard form	36
Decimals in action	38
Fractions	40
Ratio and proportion	48
Percentages	52
Converting fractions, decimals, and percentages	56
Mental maths	58
Rounding off	62
Using a calculator	64
Personal finance	66
Business finance	68

2 GEOMETRY

What is geometry?	72
Tools in geometry	74
Angles	76
Straight lines	78
Symmetry	80
Coordinates	82
Vectors	86
Translations	90
Rotations	92
Reflections	94
Enlargements	96
Scale drawings	98
Bearings	100
Constructions	102
Loci	106
Triangles	108
Constructing triangles	110
Congruent triangles	112
Area of a triangle	114
Similar triangles	117
Pythagoras' theorem	120
Quadrilaterals	122
Polygons	126
Circles	130
Circumference and diameter	132
Area of a circle	134
Angles in a circle	136
Chords and cyclic quadrilaterals	138
Tangents	140
Arcs	142
Sectors	143
Solids	144
Volumes	146
Surface area of solids	148

3 TRIGONOMETRY

What is trigonometry?	152
Working with trigonometry	153
Finding missing sides	154
Finding missing angles	156

4 ALGEBRA

What is algebra?	160
Sequences	162
Working with expressions	164
Expanding and factorizing expressions	166
Quadratic expressions	168
Formulas	169
Solving equations	172
Linear graphs	174
Simultaneous equations	178
Factorizing quadratic equations	182
The quadratic formula	184
Quadratic graphs	186
Inequalities	190

5 STATISTICS

What is statistics?	194
Collecting and organizing data	196
Bar charts	198
Pie charts	202
Line graphs	204
Averages	206
Moving Averages	210
Measuring spread	212
Histograms	216
Scatter diagrams	218

6 PROBABILITY

What is probability?	222
Expectation and reality	224
Multiple probability	226
Dependent events	228
Tree diagrams	230

Reference section	232
Glossary	244
Index	252
Acknowledgements	256

Foreword

Hello

Welcome to the wonderful world of maths. Research has shown just how important it is for a parent to be able to help a child with their education. Being able to work through homework together and enjoy a subject, particularly maths, is a vital part of a child's progress.

However, maths homework can be the cause of upset in many households. The introduction of new methods of arithmetic hasn't helped, as many parents are now simply unable to assist.

We wanted this book to guide parents through some of the methods in early arithmetic and then for them to go on to enjoy some deeper mathematics.

As a parent, I know just how important it is to be aware of when your child is struggling and equally, where they are shining. By having a greater understanding of maths, we can appreciate this even more.

Over nearly 30 years, and for nearly every single day, I have had the privilege of hearing people's very personal views about maths and arithmetic. Many weren't taught maths particularly well or in an interesting way. If you were one of those people, then we hope that this book can go some way to changing your situation and that maths, once understood, can begin to excite you as much as it does me.

CAROL VORDERMAN

Carol is the founder of her own maths school online
www.themathsfactor.com

π=**3.14**1592653589793238462643383
2795028841971693993751058209749
4459230781640628620899862803485
3421170679821480865132823066470
9384460955058223172535940812848
1117450284102701938521105559644
6229489549303819644288109756659
3344612847564823378678316527120
1909145648566923460348610454326
6482133936072602491412737245870
0660631558817488152092096282925
4091715364367892590360011330530
5488204665213841469519451160943
3057270365759591953092186117381
9326117931051185480744623799627
4956735188575272489122793818301
19491

Introduction

This book concentrates on the mathematics tackled in schools between the ages of 9 and 16. But it does so in a gripping, engaging, and visual way. Its purpose is to teach maths by stealth. It presents mathematical ideas, techniques, and procedures so that they are immediately absorbed and understood. Every spread in the book is written and presented so that the reader will exclaim, "Ah ha – now I understand!". Pupils can use it on their own; equally, it helps a parent understand and remember the subject and so, help their child. If parents too gain something in the process, then so much the better.

At the start of the new millennium I had the privilege of being the Director of **Maths Year 2000**, a celebration of mathematics and an international effort to highlight and boost awareness of the subject. It was supported by the government and Carol Vorderman was also involved. Carol championed mathematics across the British media, but is well known for her astonishingly agile ways of manipulating and working with numbers – almost as if they were her personal friends. My working, domestic, and sleeping hours are devoted to mathematics – finding out how various subtle patterns based on counting items in sophisticated structures work and how they hang together. What united us was a shared passion for mathematics and the contribution it makes to all our lives – economic, cultural, and practical.

How is it that in a world ever more dominated by numbers, mathematics – the subtle art that teases out the patterns, the harmonies, and the textures that make up the relationships between the numbers – is in danger ? I sometimes think that we are drowning in numbers.

As employees, our contribution is measured by targets, statistics, workforce percentages, and adherence to budget. As consumers, we are counted and aggregated according to every act of consumption. And in a nice subtlety, most of the products that we do consume come complete with their own personal statistics – the energy in the tin of beans and its lo (sic) salt content; the story in a newspaper and its swathe of statistics controlling and interpreting the world, developing each truth, simplifying each problem. Each minute of every hour; each hour of every day, we record and publish ever more readings from our collective life support machine. That is how we seek to understand the world, but the problem is, the more figures we get, the more truth seems to slip through our fingers.

The danger is, despite all the numbers and our increasingly numerate world, maths gets left behind. I'm sure that many think the ability to do the numbers is enough. Not so. Neither as individuals, nor collectively. Numbers are pinpricks in the fabric of mathematics, blazing within. Without them we would be condemned to total darkness. With them we gain glimpses of the sparkling treasures otherwise hidden.

This book sets out to address and solve this problem. Everyone can do maths.

BARRY LEWIS

Former President, **The Mathematical Association**.

Numbers

2 Introducing numbers

COUNTING AND NUMBERS FORM THE FOUNDATION OF MATHEMATICS.

Numbers are symbols that developed as a way to record amounts or quantities, but over centuries mathematicians have discovered ways to use and interpret numbers in order to work out new information.

What are numbers?

Numbers are basically a set of standard symbols that represent quantities – the familiar 0 to 9. In addition to these whole numbers (also called integers) there are also fractions (see pp.40–47) and decimals (see pp.38–39). Numbers can also be negative, or less than zero (see pp.30–31).

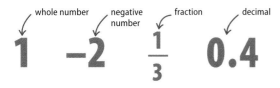

whole number   negative number   fraction   decimal

$$1 \quad -2 \quad \frac{1}{3} \quad 0.4$$

△ **Types of numbers**
Here 1 is a positive whole number and -2 is a negative number. The symbol ⅓ represents a fraction, which is one part of a whole that has been divided into three parts. A decimal is another way to express a fraction.

LOOKING CLOSER

Zero

The use of the symbol for zero is considered an important advance in the way numbers are written. Before the symbol for zero was adopted, a blank space was used in calculations. This could lead to ambiguity and made numbers easier to confuse. For example, it was difficult to distinguish between 400, 40, and 4, since they were all represented by only the number 4. The symbol zero developed from a dot first used by Indian mathematicians to act a placeholder.

◁ **Easy to read**
The zero acts as a placeholder for the "tens", which makes it easy to distinguish the single minutes.

zero is important for 24-hour timekeeping

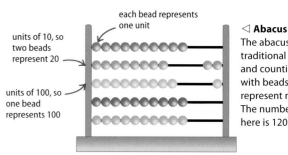

each bead represents one unit

units of 10, so two beads represent 20

units of 100, so one bead represents 100

◁ **Abacus**
The abacus is a traditional calculating and counting device with beads that represent numbers. The number shown here is 120.

▽ **First number**
One is not a prime number. It is called the "multiplicative identity", because any number multiplied by 1 gives that number as the answer.

▽ **Even prime number**
The number 2 is the only even-numbered prime number – a number that is only divisible by itself and 1 (see pp.26–27).

△ **Perfect number**
This is the smallest perfect number, which is a number that is the sum of its positive divisors (except itself). So, 1 + 2 + 3 = 6.

△ **Not the sum of squares**
The number 7 is the lowest number that cannot be represented as the sum of the squares of three whole numbers (integers).

Number symbols

Many civilizations developed their own symbols for numbers, some of which are shown below, together with our modern Hindu–Arabic number system. One of the main advantages of our modern number system is that arithmetical operations, such as multiplication and division, are much easier to do than with the more complicated older number systems.

Modern Hindu–Arabic	1	2	3	4	5	6	7	8	9	10
Mayan	•	••	•••	••••	—	⎯•⎯	⎯••⎯	⎯•••⎯	⎯••••⎯	═
Ancient Chinese	一	二	三	四	五	六	七	八	九	十
Ancient Roman	I	II	III	IV	V	VI	VII	VIII	IX	X
Ancient Egyptian			|||	⫿	⫿⫿	⫿⫿⫿	⫿⫿⫿⫿	⫿⫿⫿⫿	⫿⫿⫿⫿⫿	∩
Babylonian	𒁹	𒈫	𒐲	𒐖	𒐕	𒐗	𒐘	𒐙	𒐚	⟨

▽ **Triangular number**
This is the smallest triangular number, which is a positive whole number that is the sum of consecutive whole numbers. So, 1 + 2 = 3.

△ **Fibonacci number**
The number 8 is a cube number ($2^3 = 8$) and it is the only positive Fibonacci number (see p.163), other than 1, that is a cube.

▽ **Composite number**
The number 4 is the smallest composite number – a number that is the product of other numbers. The factors of 4 are two 2s.

△ **Highest decimal**
The number 9 is the highest single-digit whole number and the highest single-digit number in the decimal system.

▽ **Prime number**
This is the only prime number to end with a 5. A 5-sided polygon is the only shape for which the number of sides and diagonals are equal.

△ **Base number**
The Western number system is based on the number 10. It is speculated that this is because humans used their fingers and toes for counting.

Addition

NUMBERS ARE ADDED TOGETHER TO FIND THEIR TOTAL.
THIS RESULT IS CALLED THE SUM.

SEE ALSO

Subtraction   **17 ›**

Positive and negative
numbers   **30–31 ›**

Adding up

An easy way to work out the sum of two numbers is a number line. It is a group of numbers arranged in a straight line that makes it possible to count up or down. In this number line, 3 is added to 1.

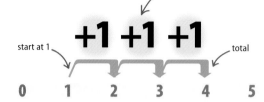

move three
steps along

start at 1   total

0   1   2   3   4   5

◁ **Use a number line**
To add 3 to 1, start at 1 and move along the line three times – first to 2, then to 3, then to 4, which is the answer.

▷ **What it means**
The result of adding 3 to the start number of 1 is 4. This means that the sum of 1 and 3 is 4.

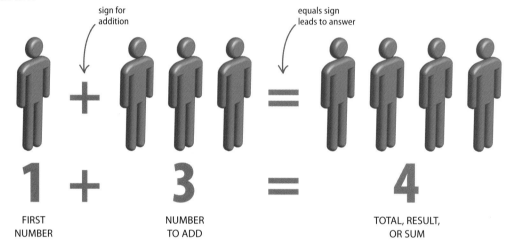

sign for
addition

equals sign
leads to answer

1   **+**   **3**   **=**   **4**

FIRST
NUMBER

NUMBER
TO ADD

TOTAL, RESULT,
OR SUM

Adding large numbers

Numbers that have two or more digits are added in vertical columns. First, add the units, then the tens, the hundreds, and so on. The sum of each column is written beneath it. If the sum has two digits, the first is carried to the next column.

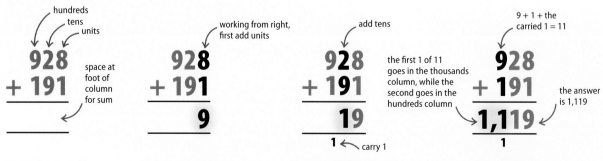

hundreds
tens
units

928
+ 191

space at
foot of
column
for sum

working from right,
first add units

928
+ 191
9

add tens

928
+ 191
19

1 ← carry 1

the first 1 of 11
goes in the thousands
column, while the
second goes in the
hundreds column

9 + 1 + the
carried 1 = 11

928
+ 191
1,119
1

the answer
is 1,119

First, the numbers
are written with their units, tens, and hundreds directly above each other.

Next, add the units 1 and 8 and write their sum of 9 in the space underneath the units column.

As the sum of the tens has two digits, write the second underneath and carry the first to the next column.

Then add the hundreds and the carried digit. As this sum has two digits, the first goes in the thousands column.

 Subtraction

A NUMBER IS SUBTRACTED FROM ANOTHER NUMBER TO
FIND WHAT IS LEFT. THIS IS KNOWN AS THE DIFFERENCE.

SEE ALSO

❮ **16** Addition

Positive and negative
numbers   **30–31** ❯

Taking away

A number line can also be used to show
how to subtract numbers. From the
first number, move back along the line
the number of places shown by the
second number. Here 3 is taken from 4.

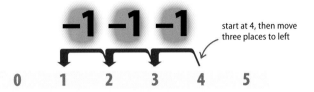

start at 4, then move
three places to left

◁ **Use a number line**
To subtract 3 from 4,
start at 4 and move
three places along the
number line, first to 3,
then 2, and then to 1.

sign for
subtraction

equals sign
leads to answer

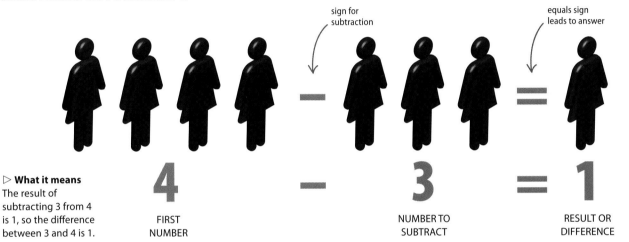

▷ **What it means**
The result of
subtracting 3 from 4
is 1, so the difference
between 3 and 4 is 1.

4
FIRST
NUMBER

–

3
NUMBER TO
SUBTRACT

=

1
RESULT OR
DIFFERENCE

Subtracting large numbers

Subtracting numbers of two or more digits is done in vertical
columns. First subtract the units, then the tens, the hundreds, and
so on. Sometimes a digit is borrowed from the next column along.

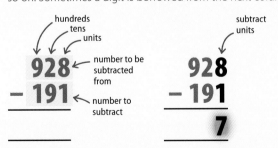

hundreds
tens
units

number to be
subtracted
from

number to
subtract

First, the numbers
are written with their
units, tens, and
hundreds directly
above each other.

subtract
units

**Next, subtract the
unit** 1 from 8, and write
their difference of 7 in
the space underneath
them.

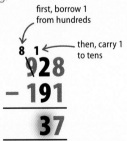

first, borrow 1
from hundreds

then, carry 1
to tens

In the tens, 9 cannot
be subtracted from 2,
so 1 is borrowed from
the hundreds, turning
9 into 8 and 2 into 12.

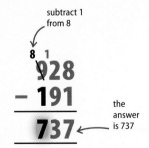

subtract 1
from 8

the
answer
is 737

In the hundreds
column, 1 is
subtracted from the
new, now lower
number of 8.

Multiplication

MULTIPLICATION INVOLVES ADDING A NUMBER TO ITSELF A NUMBER OF TIMES. THE RESULT OF MULTIPLYING NUMBERS IS CALLED THE PRODUCT.

SEE ALSO	
❮ **16–17** Addition and Subtraction	
Division	**22–25** ❯
Decimals in action	**38–39** ❯
Reference	**234–235** ❯

What is multiplication?

The second number in a multiplication sum is the number to be added to itself and the first is the number of times to add it. Here the number of rows of people is added together a number of times determined by the number of people in each row. This multiplication sum gives the total number of people in the group.

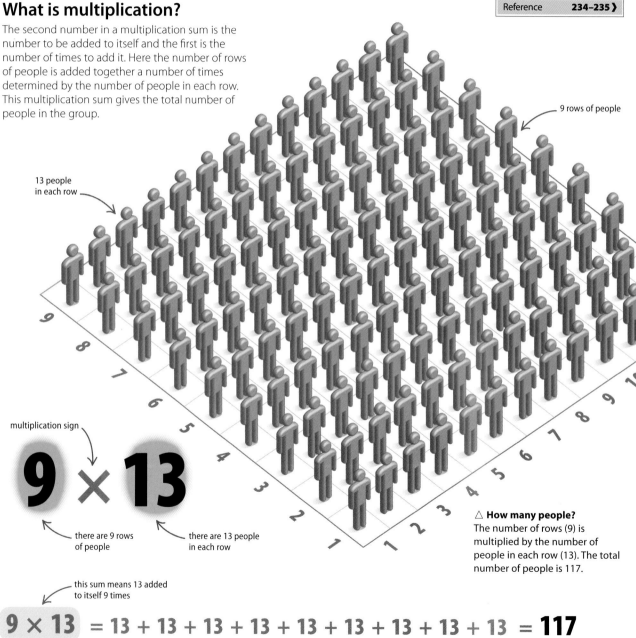

9 rows of people

13 people in each row

multiplication sign

9×13

there are 9 rows of people

there are 13 people in each row

△ **How many people?**
The number of rows (9) is multiplied by the number of people in each row (13). The total number of people is 117.

this sum means 13 added to itself 9 times

$$9 \times 13 = 13 + 13 + 13 + 13 + 13 + 13 + 13 + 13 + 13 = 117$$

product of 9 and 13 is 117

Works both ways

It does not matter which order numbers appear in a multiplication sum because the answer will be the same either way. Two methods of the same multiplication are shown here.

$$4 \times 3 = 3 + 3 + 3 + 3 = 12$$

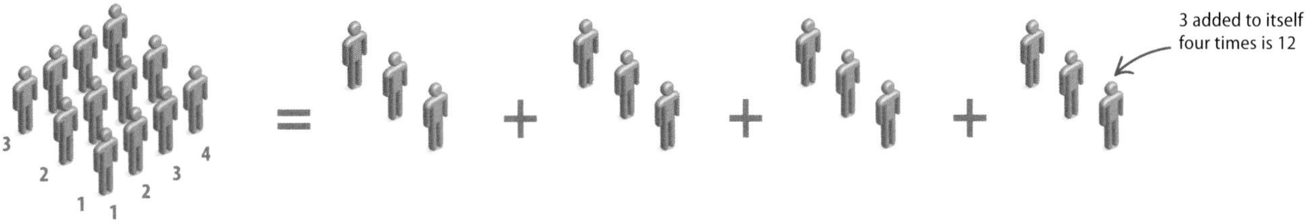

3 added to itself four times is 12

$$3 \times 4 = 4 + 4 + 4 = 12$$

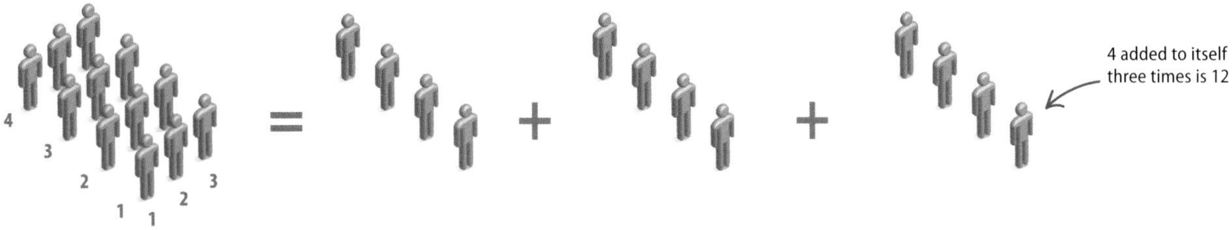

4 added to itself three times is 12

Multiplying by 10, 100, 1,000

Multiplying whole numbers by 10, 100, 1,000, and so on involves adding one zero (0), two zeroes (00), three zeroes (000), and so on to the right of the start number.

add 0 to end of start number

$$34 \times 10 = 340$$

add 00 to end of start number

$$72 \times 100 = 7,200$$

add 000 to end of start number

$$18 \times 1,000 = 18,000$$

Patterns of multiplication

There are quick ways to multiply two numbers, and these patterns of multiplication are easy to remember. The table shows patterns involved in multiplying numbers by 2, 5, 6, 9, 12, and 20.

PATTERNS OF MULTIPLICATION		
To multiply	How to do it	Example to multiply
2	add the number to itself	$2 \times 11 = 11 + 11 = 22$
5	the last digit of the number follows the pattern 5, 0, 5, 0	5, 10, 15, 20
6	multiplying 6 by any even number gives an answer that ends in the same last digit as the even number	$6 \times 12 = 72$ $6 \times 8 = 48$
9	multiply the number by 10, then subtract the number	$9 \times 7 = 10 \times 7 - 7 = 63$
12	multiply the original number first by 10, then multiply the original number by 2, and then add the two answers	$12 \times 10 = 120$ $12 \times 2 = 24$ $120 + 24 = 144$
20	multiply the number by 10 then multiply the answer by 2	$14 \times 20 =$ $14 \times 10 = 140$ $140 \times 2 = 280$

MULTIPLES

When a number is multiplied by any whole number the result (product) is called a multiple. For example, the first six multiples of the number 2 are 2, 4, 6, 8, 10, and 12. This is because $2 \times 1 = 2$, $2 \times 2 = 4$, $2 \times 3 = 6$, $2 \times 4 = 8$, $2 \times 5 = 10$, and $2 \times 6 = 12$.

MULTIPLES OF 3

$3 \times 1 = \mathbf{3}$
$3 \times 2 = \mathbf{6}$
$3 \times 3 = \mathbf{9}$
$3 \times 4 = \mathbf{12}$
$3 \times 5 = \mathbf{15}$

first five multiples of 3

MULTIPLES OF 8

$8 \times 1 = \mathbf{8}$
$8 \times 2 = \mathbf{16}$
$8 \times 3 = \mathbf{24}$
$8 \times 4 = \mathbf{32}$
$8 \times 5 = \mathbf{40}$

first five multiples of 8

MULTIPLES OF 12

$12 \times 1 = \mathbf{12}$
$12 \times 2 = \mathbf{24}$
$12 \times 3 = \mathbf{36}$
$12 \times 4 = \mathbf{48}$
$12 \times 5 = \mathbf{60}$

first five multiples of 12

Common multiples

Two or more numbers can have multiples in common. Drawing a grid, such as the one on the right, can help find the common multiples of different numbers. The smallest of these common numbers is called the lowest common multiple.

Lowest common multiple
The lowest common multiple of 3 and 8 is 24 because it is the smallest number that both multiply into

24

 multiples of 3

 multiples of 8

 multiples of 3 and 8

▷ **Finding common multiples**
Multiples of 3 and multiples of 8 are highlighted on this grid. Some multiples are common to both numbers.

1	2	3	4	5	6	7	8	9	10
11	12	13	14	15	16	17	18	19	20
21	22	23	24	25	26	27	28	29	30
31	32	33	34	35	36	37	38	39	40
41	42	43	44	45	46	47	48	49	50
51	52	53	54	55	56	57	58	59	60
61	62	63	64	65	66	67	68	69	70
71	72	73	74	75	76	77	78	79	80
81	82	83	84	85	86	87	88	89	90
91	92	93	94	95	96	97	98	99	100

Short multiplication

Multiplying a large number by a single-digit number is called short multiplication. The smaller number is placed below the larger one and aligned under the units column of the larger number.

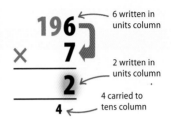

6 written in units column

2 written in units column

4 carried to tens column

To multiply 196 and 7, first multiply the units 7 and 6. The product is 42, the 4 of which is carried.

9 written in tens column

7 written in tens column

6 carried to hundreds column

Next, multiply 7 and 9, the product of which is 63. The carried 4 is added to 63 to get 67.

1 written in hundreds column

3 written in hundreds column; 1 written in thousands column

1,372 is final answer

Finally, multiply 7 and 1. Add the product (7) to the carried 6 to get 13, giving a final product of 1,372.

Long multiplication

Multiplying two numbers that both contain at least two digits is called long multiplication. The numbers are placed one above the other, in columns arranged according to their value (units, tens, hundreds, and so on).

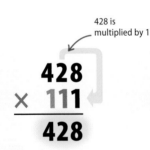

428 is multiplied by 1

First, multiply 428 by 1 in the units column. Work digit by digit from right to left so 8×1, 2×1, and then 4×1.

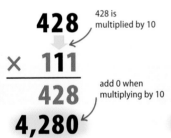

428 is multiplied by 10

add 0 when multiplying by 10

Multiply 428 by 1 in the tens column, working digit by digit. Remember to add 0 to the product when multiplying by 10.

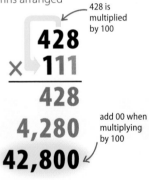

428 is multiplied by 100

add 00 when multiplying by 100

Multiply 428 by 1 in the hundreds column, digit by digit. Add 00 to the product when multiplying by 100.

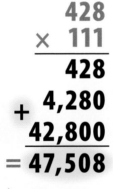

Add together the products of the three multiplications. The answer is 47,508.

Box method of multiplication

The long multiplication of 428 and 111 can be broken down into simple multiplications with the help of a table or a box. Each number is reduced to its hundreds, tens, and units, and multiplied by the other.

▷ **The final step**
Add together the nine multiplications to find the final answer.

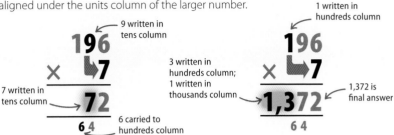

428 WRITTEN IN 100S, 10S, AND UNITS			
	400	**20**	**8**
100	400×100 $= 40{,}000$	20×100 $= 2{,}000$	8×100 $= 800$
10	400×10 $= 4{,}000$	20×10 $= 200$	8×10 $= 80$
1	400×1 $= 400$	20×1 $= 20$	8×1 $= 8$

111 WRITTEN IN 100S, 10S, AND UNITS

```
  40,000
   2,000
     800
   4,000
     200
      80
     400
      20
+      8
= 47,508
```

this is the final answer

Division

DIVISION INVOLVES FINDING OUT HOW MANY TIMES ONE NUMBER FITS INTO ANOTHER NUMBER.

SEE ALSO

⟨ **16–17** Addition and subtraction

⟨ **18–21** Multiplication

Ratio and proportion **48–51** ⟩

There are two ways to think about division. The first is sharing a number out equally (10 coins to 2 people is 5 each). The other is dividing a number into equal groups (10 coins into piles containing 2 coins each is 5 piles).

How division works

Dividing one number by another finds out how many times the second number (the divisor) fits into the first (the dividend). For example, dividing 10 by 2 finds out how many times 2 fits into 10. The result of the division is known as the quotient.

◁ **Division symbols**
There are three main symbols for division that all mean the same thing. For example, "6 divided by 3" can be expressed as 6 ÷ 3, 6/3, or $\frac{6}{3}$.

▽ **Division as sharing**
Sharing equally is one type of division. Dividing four sweets equally between two people means that each person gets the same number of sweets: two each.

4 SWEETS **÷ 2** PEOPLE **= 2** SWEETS PER PERSON

DIVIDEND
The number that is being divided or shared by another number

DIVISOR
The number that is being used to divide the dividend

LOOKING CLOSER

How division is linked to multiplication

Division is the direct opposite or "inverse" of multiplication, and the two are always connected. If you know the answer to a particular division, you can form a multiplication from it and vice versa.

◁ **Back to the beginning**
If 10 (the dividend) is divided by 2 (the divisor), the answer (the quotient) is 5. Multiplying the quotient (5) by the divisor of the original division sum (2) results in the original dividend (10).

10 ÷ 2 = 5   **5 × 2 = 10**

Another approach to division

Instead of thinking of it as sharing out a number, division can also be viewed as finding out how many groups of the second number (divisor) are contained in the first number (dividend). The division sum remains the same in both sharing and grouping.

10 SWEETS

▽ **Introducing remainders**
In this example, 10 sweets are being divided between 3 girls. However, 3 does not divide exactly into 10 – it fits 3 times with 1 left over. The amount left over from a division sum is called the remainder.

10

DIVISION

3 GIRLS

This example shows 30 footballs, which are to be divided into groups of 3:

group of three

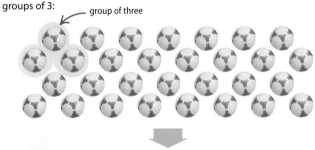

There are exactly 10 groups of 3 footballs, with no remainder, so 30 ÷ 3 = **10**.

3 SWEETS EACH

3 **3**

3 **1 REMAINING SWEET**

1

QUOTIENT
The result of the division

3 remainder 1

REMAINDER
The amount left over when one number cannot divide exactly into another

DIVISION TIPS		
A number is divisible by	**If...**	**Examples**
2	the last digit is an even number	12, 134, 5,000
3	the sum of all digits when added together is divisible by 3	18 1+8 = 9
4	the number formed by the last two digits is divisible by 4	732 32÷4 = 8
5	the last digit is 5 or 0	25, 90, 835
6	the last digit is even and the sum of its digits when added together is divisible by 3	3,426 3+4+2+6 = 15
7	no simple divisibility test	
8	the number formed by the last three digits is divisible by 8	7,536 536÷8 = 67
9	the sum of all of its digits is divisible by 9	6,831 6+8+3+1 = 18
10	the number ends in 0	30, 150, 4,270

Short division

Short division is used to divide one number (the dividend) by another whole number (the divisor) that is less than 10.

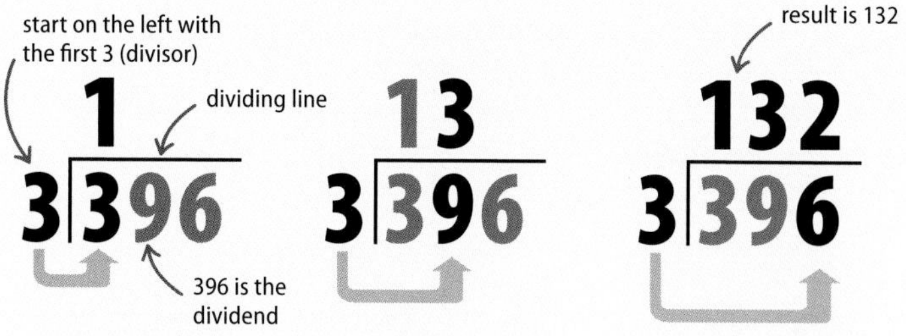

start on the left with the first 3 (divisor)

dividing line

result is 132

1

3 | **396**

396 is the dividend

Divide the first 3 into 3. It fits once exactly, so put a 1 above the dividing line, directly above the 3 of the dividend.

13

3 | **396**

Move to the next column and divide 3 into 9. It fits three times exactly, so put a 3 directly above the 9 of the dividend.

132

3 | **396**

Divide 3 into 6, the last digit of the dividend. It goes twice exactly, so put a 2 directly above the 6 of the dividend.

Carrying numbers

When the result of a division gives a whole number and a remainder, the remainder can be carried over to the next digit of the dividend.

start on the left

divisor

5 | **2,765**

2,765 is the dividend

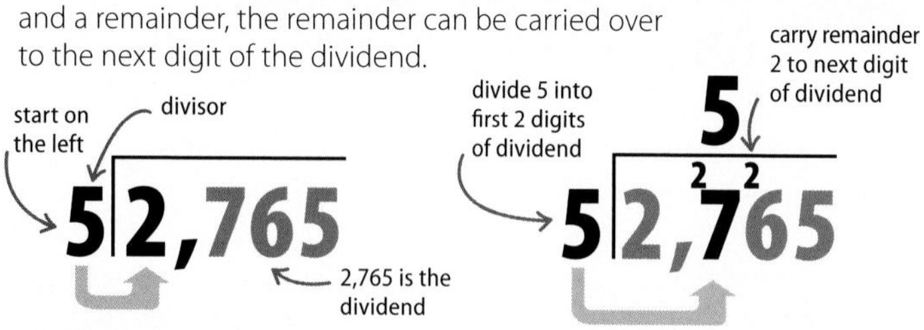

Start with number 5. It does not divide into 2 as it is a larger number. Instead, 5 will need to be divided into the first two digits of the dividend.

divide 5 into first 2 digits of dividend

carry remainder 2 to next digit of dividend

5

5 | **2,²7⁶5**

Divide 5 into 27. The result is 5 with a remainder of 2. Put 5 directly above the 7 and carry the remainder.

carry remainder 1 to next digit of dividend

55

5 | **2,²7⁶¹5**

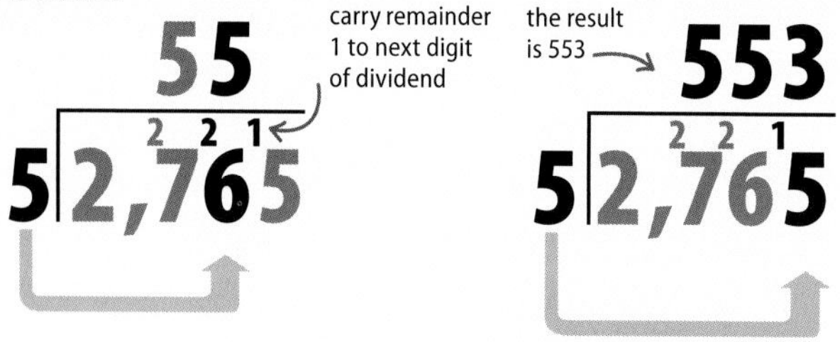

Divide 5 into 26. The result is 5 with a remainder of 1. Put 5 directly above the 6 and carry the remainder 1 to the next digit of the dividend.

the result is 553

553

5 | **2,²7⁶¹5**

Divide 5 into 15. It fits three times exactly, so put 3 above the dividing line, directly above the final 5 of the dividend.

Converting remainders

When one number will not divide exactly into another, the answer has a remainder. Remainders can be converted into decimals, as shown below.

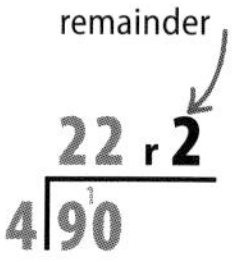

remainder

22 r 2

4 | **90**

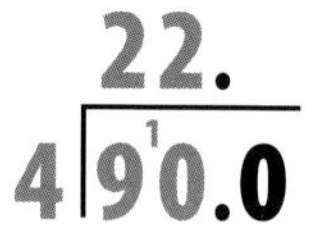

22.

4 | **9ⁱ0.0**

Remove the remainder, 2 in this case, leaving 22. Add a decimal point above and below the dividing line. Next, add a zero to the dividend after the decimal point.

22.

4 | **9ⁱ0.²0**

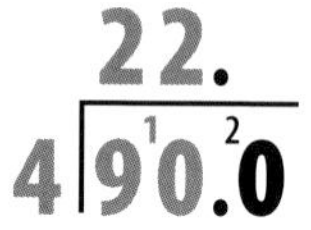

Carry the remainder (2) from above the dividing line to below the line and put it in front of the new zero.

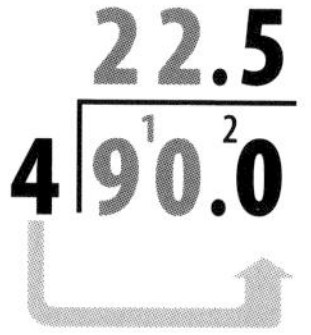

22.5

4 | **9ⁱ0.²0**

Divide 4 into 20. It goes 5 times exactly, so put a 5 directly above the zero of the dividend and after the decimal point.

Making division simpler

To make a division easier, sometimes the divisor can be split into factors. This means that a number of simpler divisions can be done.

816÷6

divisor is 6, which is 2×3. Splitting 6 into 2 and 3 simplifies the sum

result is 136

816÷2 = 408 ▶ **408÷3 = 136**

divide by first factor of divisor

divide by second factor of divisor

This method of splitting the divisor into factors can also be used for more difficult divisions.

405÷15

splitting 15 into 5 and 3, which multiply to make 15, simplifies the sum

result is 27

405÷5 = 81 ▶ **81÷3 = 27**

divide by first factor of divisor

divide result by second factor of divisor

Long division

Long division is usually used when the divisor is at least two digits long and the dividend is at least 3 digits long. Unlike short division, all the workings out are written out in full below the dividing line. Multiplication is used for finding remainders. A long division sum is presented in the example on the right.

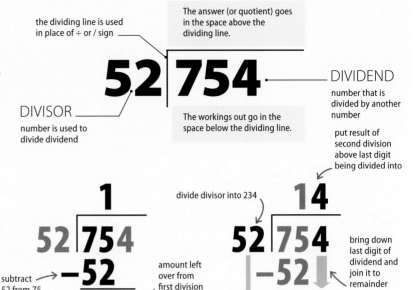

the dividing line is used in place of ÷ or / sign

The answer (or quotient) goes in the space above the dividing line.

DIVIDEND
number that is divided by another number

DIVISOR
number is used to divide dividend

The workings out go in the space below the dividing line.

put result of second division above last digit being divided into

result is 1

divide divisor into first two digits of dividend

Begin by dividing the divisor into the first two digits of the dividend. 52 fits into 75 once, so put a 1 above the dividing line, aligning it with the last digit of the number being divided.

subtract 52 from 75

amount left over from first division

Work out the first remainder. The divisor 52 does not divide into 75 exactly. To work out the amount left over (the remainder), subtract 52 from 75. The result is 23.

divide divisor into 234

bring down last digit of dividend and join it to remainder

Now, bring down the last digit of the dividend and place it next to the remainder to form 234. Next, divide 234 by 52. It goes four times, so put a 4 next to the 1.

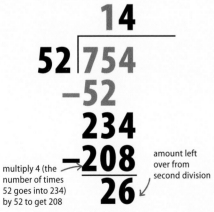

multiply 4 (the number of times 52 goes into 234) by 52 to get 208

amount left over from second division

Work out the second remainder. The divisor, 52, does not divide into 234 exactly. To find the remainder, multiply 4 by 52 to make 208. Subtract 208 from 234, leaving 26.

add a decimal point then a zero

bring down zero and join it to remainder

There are no more whole numbers to bring down, so add a decimal point after the dividend and a zero after it. Bring down the zero and join it to the remainder 26 to form 260.

add decimal point above other one

put result of last sum after decimal point

Put a decimal point after the 14. Next, divide 260 by 52, which goes five times exactly. Put a 5 above the dividing line, aligned with the new zero in the dividend.

11 Prime numbers

ANY WHOLE NUMBER LARGER THAN 1 THAT CANNOT BE DIVIDED
BY ANY OTHER NUMBER EXCEPT FOR ITSELF AND 1.

SEE ALSO
❮ 18–21 Multiplication
❮ 22–25 Division

Introducing prime numbers

Over 2,000 years ago, the Ancient Greek mathematician Euclid noted that
some numbers are only divisible by 1 or the number itself. These numbers
are known as prime numbers. A number that is not a prime is called a
composite – it can be arrived at, or composed, by multiplying together
smaller prime numbers, which are known as its prime factors.

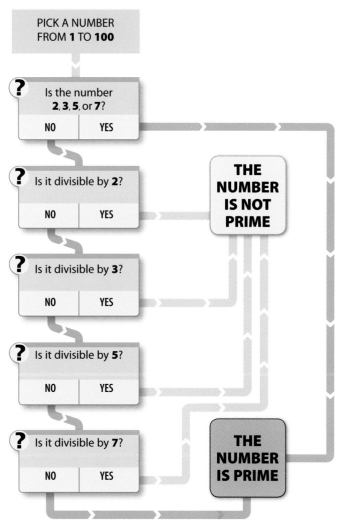

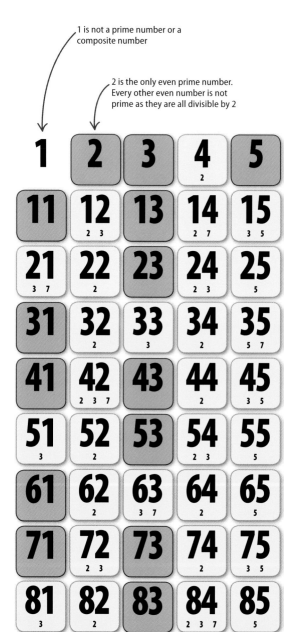

1 is not a prime number or a
composite number

2 is the only even prime number.
Every other even number is not
prime as they are all divisible by 2

△ **Is a number prime?**
This flowchart can be used to determine whether a
number between 1 and 100 is prime by checking if it
is divisible by any of the primes 2, 3, 5, and 7.

▷ **First 100 numbers**
This table shows the prime
numbers among the first
100 whole numbers.

KEY

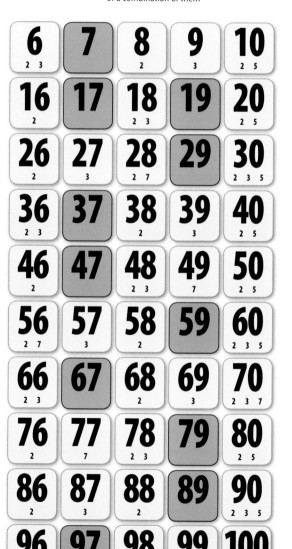

17
Prime number
A blue box indicates that the number is prime. It has no factors other than 1 and itself.

42
2 3 7
Composite number
A yellow box denotes a composite number, which means that it is divisible by more than 1 and itself.

smaller numbers show whether the number is divisible by 2, 3, 5, or 7, or a combination of them

Prime factors

Every number is either a prime or the result of multiplying together prime numbers. Prime factorization is the process of breaking down a composite number into the prime numbers that it is made up of. These are known as its prime factors.

prime factor ⟶   ⟵ remaining factor

$$30 = 5 \times 6$$

To find the prime factors of 30, find the largest prime number that divides into 30, which is 5. The remaining factor is 6 (5 x 6 = 30), which needs to be broken down into prime numbers.

largest prime factor

$$6 = 3 \times 2$$

Next, take the remaining factor and find the largest prime number that divides into it, and any smaller prime numbers. In this case, the prime numbers that divide into 6 are 3 and 2.

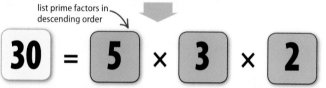

list prime factors in descending order

$$30 = 5 \times 3 \times 2$$

It is now possible to see that 30 is the product of multiplying together the prime numbers 5, 3, and 2. Therefore, the prime factors of 30 are 5, 3, and, 2.

REAL WORLD

Encryption

Many transactions in banks and shops rely on the Internet and other communications systems. To protect the information, it is coded using a number that is the product of just two huge primes. The security relies on the fact that no "eavesdropper" can factorize the number because its factors are so large.

▷ **Data protection**
To provide constant security, mathematicians relentlessly hunt for ever bigger primes.

fldjhg83asldkfdslkfjour523ijwli
eorit84wodfpflciry38s0x8b6lkj
qpeoith73kdicuvyebdkciurmol
wpeodikrucnyr83iowp7uhjwm
kdieolekdori**password**qe8ki
mdkdoritut6483kednffkeoskeo
kdieujr83iowplwqpwo98irkldil
ieow98mqloapkijuhrnmeuidy6
woqp90jqiuke4lmicunejwkiuyj

Units of measurement

UNITS OF MEASUREMENT ARE STANDARD SIZES USED TO MEASURE TIME, MASS, AND LENGTH.

SEE ALSO

Volumes	146–147 ⟩
Formulas	169–171 ⟩
Reference	234–237 ⟩

Basic units

A unit is any agreed or standardized measurement of size. This allows quantities to be accurately measured. There are three basic units: time, weight (including mass), and length.

LOOKING CLOSER

Distance

The distance is the amount of space between two points. It expresses length, but is also used to describe a journey, which is not always the most direct route between two points.

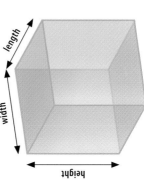

plane flies set distance between two cities

distance between cities A and B

△ Time
Time is measured in milliseconds, seconds, minutes, hours, days, weeks, months, and years. Different countries and cultures may have calendars which start a new year at a different time.

△ Weight and mass
Weight is how heavy something is in relation to the force of gravity acting upon it. Mass is the amount of matter that makes up the object. Both are measured in the same units, such as grams and kilograms, or ounces and pounds.

these three units are heavier

these two units are lighter

this is the height of the building

this is the width of the building

this is the length of the building

△ Length
Length is how long something is. It is measured in centimetres, metres, and kilometres in the metric system, or in inches, feet, yards, and miles in the imperial system (see pp.234–237).

Compound measures

A compound unit is made up of more than one of the basic units, including using the same unit repeatedly. Examples include area, volume, speed, and density.

▽ Area
Area is measured in squared units. The area of a square is the product of its length and width; if they were both measured in metres (m), its area would be m × m, which is written as m².

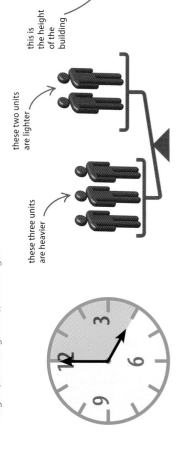

$$\text{area} = \text{length} \times \text{width}$$

area is made up of two of the same units, as width is also a length

▽ Volume
Volume is measured in cubed units. The volume of a cuboid is the product of its height, width, and length; if they were all measured in metres (m), its area would be m × m × m, or m³.

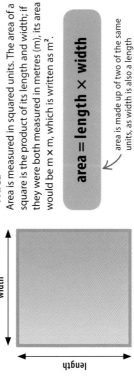

$$\text{volume} = \text{length} \times \text{width} \times \text{height}$$

volume is a compound of three of the same units, as width and height are technically lengths

Speed

Speed measures the distance (length) travelled in a given time. This means that the formula for measuring speed is length ÷ time. If this is measured in kilometres and hours, the unit for speed will be km/h.

$$\text{Speed} = \frac{\text{distance}}{\text{time}}$$

△ Speed formula triangle
The relationships between speed, distance, and time can be shown in a triangle. The position of each unit in the triangle indicates how to use the other two measurements to calculate that unit.

this line acts as a multiplication sign

this line acts as a division sign

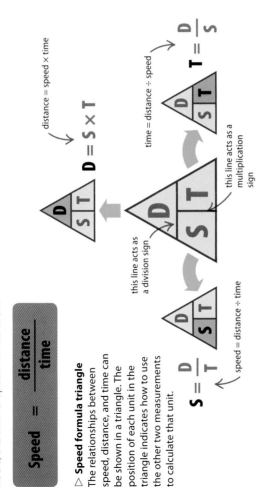

$$D = S \times T$$
distance = speed × time

$$T = \frac{D}{S}$$
time = distance ÷ speed

$$S = \frac{D}{T}$$
speed = distance ÷ time

▷ Finding speed
A van travels 20km in 20 minutes. From this information its speed in km/h can be found.

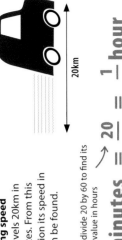

20km

First, convert the minutes into hours. To convert minutes into hours, divide them by 60, then cancel the fraction – divide the top and bottom numbers by 20. This gives an answer of $^1/_3$ hour.

divide 20 by 60 to find its value in hours

$$20 \text{ minutes} = \frac{20}{60} = \frac{1}{3} \text{ hour}$$

Then, substitute the values for distance and time into the formula for speed. Divide the distance (20km) by the time ($^1/_3$ hour) to find the speed, in this case 60 km/h.

distance is 20km

$$S = \frac{D}{T} = 60\text{km/h}$$

time is $^1/_3$ hour

Density

Density measures how much matter is packed into a given volume of a substance. It involves two units – mass and volume. The formula for measuring density is mass ÷ volume. If this is measured in grams and centimetres, the unit for density will be g/cm³.

$$\text{Density} = \frac{\text{mass}}{\text{volume}}$$

△ Density formula triangle
The relationships between density, mass, and volume can be shown in a triangle. The position of each unit of measurement in the triangle shows how to calculate that unit using the other two measurements.

this line acts as a multiplication sign

this line acts as a division sign

$$M = D \times V$$
mass = density × volume

$$V = \frac{M}{D}$$
volume = mass ÷ density

$$D = \frac{M}{V}$$
density = mass ÷ volume

▷ Finding volume
Lead has a density of 0.0113kg/cm³. With this measurement, the volume of a lead weight that has a mass of 0.5kg can be found.

density of lead is constant, regardless of mass

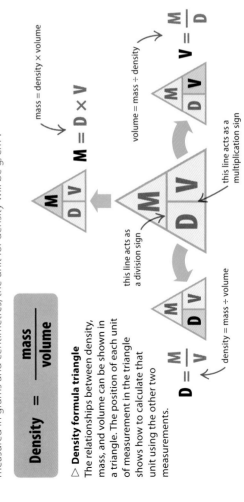

0.5kg

△ Using the formula
Substitute the values for mass and density into the formula for volume. Divide the mass (0.5kg) by the density (0.0113kg/cm³) to find the volume, in this case 44.25cm³.

mass is 0.5kg

$$V = \frac{M}{D} = 44.25\text{cm}^3$$

density is 0.0113kg/cm³

Positive and negative numbers

A POSITIVE NUMBER IS A NUMBER THAT IS MORE THAN ZERO
(NOUGHT), WHILE A NEGATIVE NUMBER IS LESS THAN ZERO.

A positive number is shown by a plus sign (+), or has no sign in front of it.
If a number is negative, it has a minus sign (–) in front of it.

SEE ALSO
❰ **14–15** Introducing numbers
❰ **16–17** Addition and subtraction

Why use positives and negatives?

Positive numbers are used when an amount is counted up from
zero, and negative numbers when it is counted down from
zero. For example, if a bank account has money in it, it is a
positive amount of money, but if the account is overdrawn,
the amount of money in the account is negative.

negative number

number line continues forever

−5   **−4**   **−3**   **−2**

Adding and subtracting positives and negatives

Use a number line to add and subtract positive and negative numbers. Find the first
number on the line and then move the number of steps shown by the second
number. Move right for addition and left for subtraction.

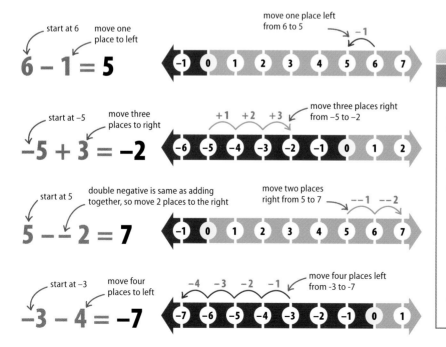

start at 6 move one place to left

move one place left from 6 to 5 −1

$$6 - 1 = 5$$

-1 0 1 2 3 4 5 6 7

start at −5 move three places to right

move three places right from −5 to −2 +1 +2 +3

$$-5 + 3 = -2$$

-6 -5 -4 -3 -2 -1 0 1 2

start at 5 double negative is same as adding together, so move 2 places to the right

move two places right from 5 to 7 − −1 − −2

$$5 - -2 = 7$$

-1 0 1 2 3 4 5 6 7

start at −3 move four places to left

move four places left from -3 to -7 −4 −3 −2 −1

$$-3 - 4 = -7$$

-7 -6 -5 -4 -3 -2 -1 0 1

LOOKING CLOSER

Double negatives

If a negative or minus number is
subtracted from a positive number, it
creates a double negative. As the first
negative is cancelled out by the second
negative, the result is always a positive;
for example 5 minus −2 is the same as
adding 2 to 5.

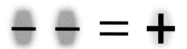

△ **Like signs equal a positive**
If any two like signs appear together,
the result is always positive. The result is
negative with two unlike signs together.

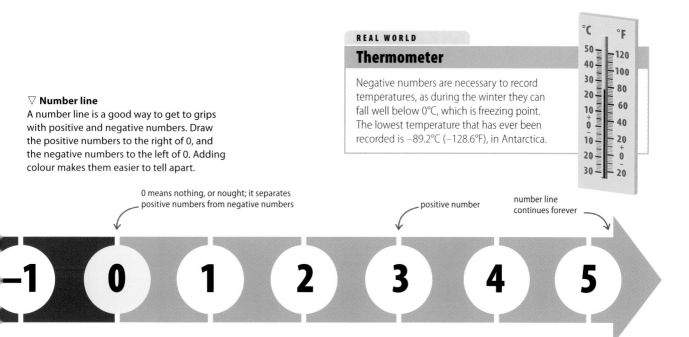

▽ **Number line**
A number line is a good way to get to grips with positive and negative numbers. Draw the positive numbers to the right of 0, and the negative numbers to the left of 0. Adding colour makes them easier to tell apart.

REAL WORLD

Thermometer

Negative numbers are necessary to record temperatures, as during the winter they can fall well below 0°C, which is freezing point. The lowest temperature that has ever been recorded is −89.2°C (−128.6°F), in Antarctica.

0 means nothing, or nought; it separates positive numbers from negative numbers

positive number

number line continues forever

−1   0   1   2   3   4   5

Multiplying and dividing

To multiply or divide any two numbers, do the sum ignoring whether they are positive or negative, then work out if the answer is positive or negative using the diagram on the right.

$2 \times 4 = 8$

8 is positive because
+ × + = +

$-1 \times 6 = -6$

−6 is negative because
− × + = −

$-4 \div 2 = -2$

−2 is negative because
− ÷ + = −

$-2 \times 4 = -8$

−8 is negative because
− × + = −

$-2 \times -4 = 8$

8 is positive because
− × − = +

$-10 \div -2 = 5$

5 is positive because
− ÷ − = +

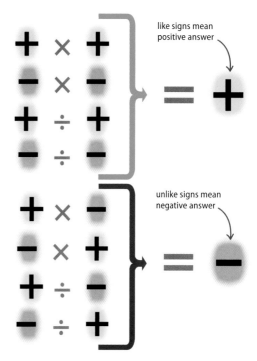

like signs mean positive answer

$+ \times + $
$- \times - $
$+ \div + $
$- \div - $
$= +$

unlike signs mean negative answer

$+ \times - $
$- \times + $
$+ \div - $
$- \div + $
$= -$

△ **Positive or negative answer**
The sign in the answer depends on whether the signs of the numbers are alike or not.

Powers and roots

A POWER IS THE NUMBER OF TIMES A NUMBER IS MULTIPLIED BY ITSELF. THE ROOT OF A NUMBER IS A NUMBER WHICH, MULTIPLIED BY ITSELF, EQUALS THE ORIGINAL NUMBER.

SEE ALSO
❮ **18–21** Multiplication
❮ **22–25** Division
Standard form **36–37** ❯
Using a calculator **64–65** ❯

Introducing powers

A power is the number of times a number is multiplied by itself. This is indicated as a smaller number positioned to the right above the number. Multiplying a number by itself once is described as "squaring" the number; multiplying a number by itself twice is described as "cubing" the number.

$$5^4$$

← this is the power, which shows how many times to multiply the number (5^4 means $5 \times 5 \times 5 \times 5$)

← this is the number that the power relates to

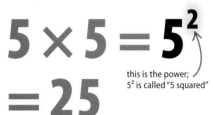

$$5 \times 5 = 5^2$$
$$= 25$$

this is the power; 5^2 is called "5 squared"

△ **The square of a number**
Multiplying a number by itself gives the square of the number. The power for a square number is ^{2}, for example 5^2, which means there are 2 lots of 5 or 5×5.

▷ **Squared number**
This image shows how many units make up 5^2. There are 5 rows, each with 5 units – so $5 \times 5 = 25$.

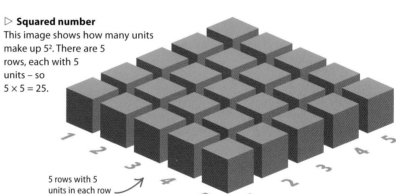

5 rows with 5 units in each row ↗

$$5 \times 5 \times 5 = 5^3$$
$$= 125$$

this is the power; 5^3 is called "5 cubed"

△ **The cube of a number**
Multiplying a number by itself twice gives its cube. The power for a cube number is ^{3}, for example 5^3, which means there are 3 lots of 5: $5 \times 5 \times 5$.

5 vertical rows ↗

▷ **Cubed number**
This image shows how many units make up 5^3. There are 5 horizontal rows and 5 vertical rows, each with 5 units in each one, so $5 \times 5 \times 5 = 125$.

5 horizontal rows ↗

5 blocks of units ←

Square roots and cube roots

A square root is a number which, multiplied by itself once, equals a given number. For example, one square root of 4 is 2, as $2 \times 2 = 4$. Another square root is −2, as $(-2) \times (-2) = 4$. A cube root is a number which, multiplied by itself twice, equals a given number. For example, the cube root of 27 is 3, as $3 \times 3 \times 3 = 27$.

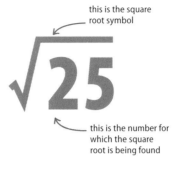

this is the square root symbol

this is the number for which the square root is being found

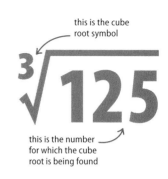

this is the cube root symbol

this is the number for which the cube root is being found

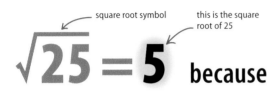

square root symbol

this is the square root of 25

because $5 \times 5 = 25$

25 is 5^2

△ The square root of a number
The square root of a number is the number which, when squared (multiplied by itself), equals the number under the square root sign.

cube root symbol

this is the cube root of 125

$\sqrt[3]{125} = 5$ because $5 \times 5 \times 5 = 125$

125 is 5^3

△ The cube root of a number
The cube root of a number is the number which, when cubed (multiplied by itself twice), equals the number under the cube root sign.

COMMON SQUARE ROOTS		
Square root	Answer	Why?
1	1	Because $1 \times 1 = 1$
4	2	Because $2 \times 2 = 4$
9	3	Because $3 \times 3 = 9$
16	4	Because $4 \times 4 = 16$
25	5	Because $5 \times 5 = 25$
36	6	Because $6 \times 6 = 36$
49	7	Because $7 \times 7 = 49$
64	8	Because $8 \times 8 = 64$
81	9	Because $9 \times 9 = 81$
100	10	Because $10 \times 10 = 100$
121	11	Because $11 \times 11 = 121$
144	12	Because $12 \times 12 = 144$
169	13	Because $13 \times 13 = 169$

LOOKING CLOSER

Using a calculator

Calculators can be used to find powers and square roots. Most calculators have buttons to square and cube numbers, buttons to find square roots and cube roots, and an exponent button, which allows them to raise numbers to any power.

△ Exponent
This button allows any number to be raised to any power.

$3^5 =$ [3] [X^y] [5]
$= 243$

◁ Using exponents
First enter the number to be raised to a power, then press the exponent button, then enter the power required.

△ Square root
This button allows the square root of any number to be found.

$25 =$ [√] [25]
$= 5$

◁ Using square roots
On most calculators, find the square root of a number by pressing the square root button first and then entering the number.

Multiplying powers of the same number

To multiply powers that have the same number simply add the powers. The power of the answer is the sum of the powers that are being multiplied.

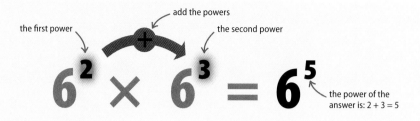

add the powers

the first power

the second power

$$6^2 \times 6^3 = 6^5$$

the power of the answer is: $2 + 3 = 5$

because

▷ **Writing it out**
Writing out what each of these powers represents shows why powers are added together to multiply them.

$$(6 \times 6) \times (6 \times 6 \times 6) = 6 \times 6 \times 6 \times 6 \times 6$$

6^2 is 6×6 6^3 is $6 \times 6 \times 6$ $6 \times 6 \times 6 \times 6 \times 6$ is 6^5

Dividing powers of the same number

To divide powers of the same number, subtract the second power from the first. The power of the answer is the difference between the first and second powers.

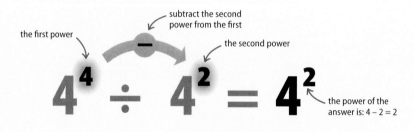

subtract the second power from the first

the first power

the second power

$$4^4 \div 4^2 = 4^2$$

the power of the answer is: $4 - 2 = 2$

because

▷ **Writing it out**
Writing out the division of the powers as a fraction and then cancelling the fraction shows why powers to be divided can simply be subtracted.

4^4 is $4 \times 4 \times 4 \times 4$

$$\frac{4 \times 4 \times 4 \times 4}{4 \times 4} \Rightarrow \frac{\cancel{4} \times \cancel{4} \times 4 \times 4}{\cancel{4} \times \cancel{4}} = 4 \times 4$$

4^2 is 4×4

cancel the fraction to its simplest terms

4×4 is 4^2

LOOKING CLOSER

Zero power

Any number raised to the power 0 is equal to 1. Dividing two equal powers of the same number gives a power of 0, and therefore the answer 1. These rules only apply when dealing with powers of the same number.

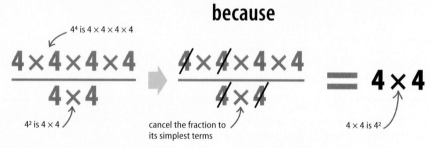

the first power

the second power

the power of the answer is: $3 - 3 = 0$

$$8^3 \div 8^3 = 8^0 = 1$$

any number to the power $0 = 1$

because

▷ **Writing it out**
Writing out the division of two equal powers makes it clear why any number to the power 0 is always equal to 1.

8^3 is $8 \times 8 \times 8$

$$\frac{8 \times 8 \times 8}{8 \times 8 \times 8} = \frac{512}{512} = 1$$

any number divided by itself $= 1$

Finding a square root by estimation

It is possible to find a square root through estimation, by choosing a number to multiply by itself, working out the answer, and then altering the number depending on whether the answer needs to be higher or lower.

$$\sqrt{32} = ?$$

$\sqrt{25} = 5$ and $\sqrt{36} = 6$, so the answer must be somewhere between 5 and 6. Start with the midpoint between the two, 5.5:

$5.5 \times 5.5 = 30.25$ — Too low

$5.75 \times 5.75 = 33.0625$ — Too high

$5.65 \times 5.65 = 31.9225$ — Too low

$5.66 \times 5.66 = \mathbf{32.0356}$

the square root of 32 is approximately 5.66 — this would round down to 32

$$\sqrt{1,000} = ?$$

$\sqrt{1,600} = 40$ and $\sqrt{900} = 30$, so the answer must be between 40 and 30. 1,000 is closer to 900 than 1,600, so start with a number closer to 30, such as 32:

$32 \times 32 = 1,024$ — Too high

$31 \times 31 = 961$ — Too low

$31.5 \times 31.5 = 992.25$ — Too low

$31.6 \times 31.6 = 998.56$ — Too low

$31.65 \times 31.65 = 1,001.72$ — Too high

$31.62 \times 31.62 = \mathbf{999.8244}$

the square root of 1,000 is approximately 31.62 — this would round up to 1,000 as the nearest whole number

Finding a cube root by estimation

Cube roots of numbers can also be estimated without a calculator. Use round numbers to start with, then use these answers to get closer to the final answer.

$$\sqrt[3]{32} = ?$$

$3 \times 3 \times 3 = 27$ and $4 \times 4 \times 4 = 64$, so the answer is somewhere between 3 and 4. Start with the midpoint between the two, 3.5:

$3.5 \times 3.5 \times 3.5 = 42.875$ — Too high

$3.3 \times 3.3 \times 3.3 = 35.937$ — Too high

$3.1 \times 3.1 \times 3.1 = 29.791$ — Too low

$3.2 \times 3.2 \times 3.2 = 32.768$ — Too high

$3.18 \times 3.18 \times 3.18 = \mathbf{32.157432}$

the cube root of 32 is approximately 3.18 — this would be 32.2 to 1 decimal place

$$\sqrt[3]{800} = ?$$

$9 \times 9 \times 9 = 729$ and $10 \times 10 \times 10 = 1,000$, so the answer is somewhere between 9 and 10. 800 is closer to 729 than 1000, so start with a number closer to 9, such as 9.1:

$9.1 \times 9.1 \times 9.1 = 753.571$ — Too low

$9.3 \times 9.3 \times 9.3 = 804.357$ — Too high

$9.27 \times 9.27 \times 9.27 = 796.5979$ — Too low

$9.28 \times 9.28 \times 9.28 = 799.1787$ — Very close

$9.284 \times 9.284 \times 9.284 = \mathbf{800.2126}$

the cube root of 32 is approximately 9.284 — this would round down to 800

Standard Form

STANDARD FORM IS A CONVENIENT WAY OF WRITING VERY
LARGE AND VERY SMALL NUMBERS.

SEE ALSO
❰ **18–21** Multiplication
❰ **22–25** Division
❰ **32–35** Powers and roots

Introducing standard form

Standard form makes very large or very small numbers
easier to understand by showing them as a number
multiplied by a power of 10. This is useful because the
size of the power of 10 makes it possible to get an
instant impression of how big the number really is.

this is the
power of 10

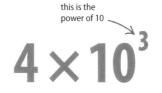

$$4 \times 10^3$$

◁ **Using standard form**
This is how 4,000 is written as
standard form – it shows that the
decimal place for the number
represented, 4,000, is 3 places to
the right of 4.

How to write a number in standard form

To write a number in standard form, work out how many places the
decimal point must move to form a number between 1 and 10. If the
number does not have a decimal point, add one after its final digit.

▷ **Take a number**
Standard form is usually used
for very large or very small
numbers.

very large number

1,230,000

very small number

0.0006

▷ **Add the decimal point**
Identify the position of the
decimal point if there is one.
Add a decimal point at the
end of the number, if it does
not already have one.

add decimal point

1,230,000.

decimal point is already here

0.0006

▷ **Move the decimal point**
Move along the number
and count how many places
the decimal point must move
to form a number between
1 and 10.

6 5 4 3 2 1

1,230,000.

the decimal point moves 6
places to the left

1 2 3 4

0.0006

the decimal point moves
4 places to the right

▷ **Write as standard form**
The number between 1 and
10 is multiplied by 10, and the
small number, the "power" of
10, is found by counting how
many places the decimal
point moved to create the first
number.

the power is 6 because the decimal point
moved six places; the power is positive because
the decimal point moved to the left

$$1.23 \times 10^6$$

the first number must
always be between 1
and 10

the power is negative
because the decimal point
moved to the right

$$6 \times 10^{-4}$$

the power is 4 because
the decimal point moved
four places

Standard form in action

Sometimes it is difficult to compare how large or small numbers are because of the number of digits they contain. Standard form makes this easier.

The mass of Earth is 5,974,200,000,000,000,000,000,000 kg

24 23 22 21 20 19 18 17 16 15 14 13 12 11 10 9 8 7 6 5 4 3 2 1

5,974,200,000,000,000,000,000,000.0 kg

The decimal point moves **24 places** to the left.

The mass of the planet Mars is

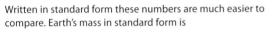

23 22 21 20 19 18 17 16 15 14 13 12 11 10 9 8 7 6 5 4 3 2 1

641,910,000,000,000,000,000,000.0 kg

The decimal point moves **23 places** to the left.

Written in standard form these numbers are much easier to compare. Earth's mass in standard form is

$$5.9742 \times 10^{24} \text{ kg}$$

The mass of Mars in standard form is

$$6.4191 \times 10^{23} \text{ kg}$$

▷ **Comparing planet mass**
It is immediately evident that the mass of the Earth is bigger than the mass of Mars, as 10^{24} is 10 times larger than 10^{23}.

EXAMPLES OF STANDARD FORM		
Example	**Decimal form**	**Standard form**
Weight of the Moon	73,600,000,000,000,000,000,000 kg	7.36×10^{22} kg
Humans on Earth	6,800,000,000	6.8×10^{9}
Speed of light	300,000,000 m/sec	3×10^{8} m/sec
Distance of the Moon from the Earth	384,000 km	3.8×10^{5} km
Weight of the Empire State building	365,000 tons	3.65×10^{5} tons
Distance around the Equator	40,075 km	4×10^{4} km
Height of Mount Everest	8,850 m	8.850×10^{3} m
Speed of a bullet	710 m/sec	7.1×10^{2} m/sec
Speed of a snail	0.001 m/sec	1×10^{-3} m/sec
Width of a red blood cell	0.00067 cm	6.7×10^{-4} cm
Length of a virus	0.000 000 009 cm	9×10^{-9} cm
Weight of a dust particle	0.000 000 000 753 kg	7.53×10^{-10} kg

LOOKING CLOSER

Standard form and calculators

The exponent button on a calculator allows a number to be raised to any power. Calculators give very large answers in standard form.

△ **Exponent button**
This calculator button allows any number to be raised to any power.

Using the exponent button:

4×10^{2} is entered by pressing

On some calculators, answers appear in standard form:

$$1234567 \times 89101112 =$$
$$1.100012925 \times 10^{14}$$

so the answer is approximately 110,001,292,500,000

Decimals

NUMBERS WRITTEN IN DECIMAL FORM ARE CALLED DECIMAL
NUMBERS OR, MORE SIMPLY, DECIMALS.

SEE ALSO

❰ **18–21** Multiplication
❰ **22–25** Division
Using a
calculator **64–65**❱

Decimal numbers

In a decimal number, the digits to the left of the decimal point are whole
numbers. The digits to the right of the decimal point are not whole numbers.
The first digit to the right of the decimal point represents tenths, the second
hundredths, and so on. These are called fractional parts.

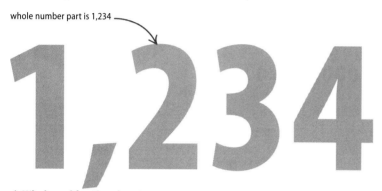

whole number part is 1,234

fractional part is 56

decimal point separates the whole
numbers (on the left) from the
fractional numbers (on the right)

△ **Whole and fractional parts**
The whole numbers represent – moving left from the decimal
point – units, tens, hundreds, and thousands. The fractional
numbers – moving right from the decimal place – are tenths
then hundredths.

Multiplication

To multiply decimals, first remove the decimal point. Then perform a long
multiplication of the two numbers, before adding the decimal point back in
to the answer. Here 1.9 (a decimal) is multiplied by 7 (a whole number).

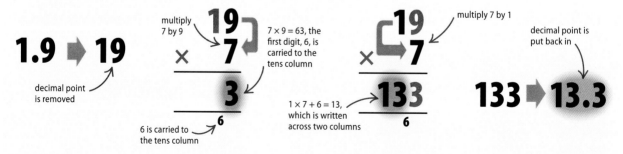

multiply
7 by 9

$7 \times 9 = 63$, the
first digit, 6, is
carried to the
tens column

6 is carried to
the tens column

$1 \times 7 + 6 = 13$,
which is written
across two columns

multiply 7 by 1

decimal point is
put back in

decimal point
is removed

First, remove any decimal
points, so that both
numbers can be treated
as whole numbers.

▶ **Then multiply** the two
numbers, starting in the
units column. Carry units to
the tens if necessary.

▶ **Next multiply the tens.** The
product is 7, which, added to
the carried 6, makes 13. Write
this across two columns.

▶ **Finally, count** the decimal
digits in the original numbers
– there is 1. The answer will
also have 1 decimal digit.

DIVISION

Dividing one number by another often gives a decimal answer. Sometimes it is easier to turn decimals into whole numbers before dividing them.

Short division with decimals

Many numbers do not divide into each other exactly. If this is the case, a decimal point is added to the number being divided, and zeros are added after the point until the division is solved. Here 6 is divided by 8.

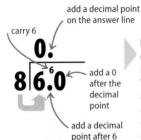

add a decimal point on the answer line

Both numbers are whole. As 8 will not divide into 6, put in a decimal point with a 0 after it and carry the 6.

carry 6

add a 0 after the decimal point

add a decimal point after 6

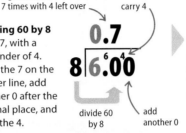

8 goes into 60 7 times with 4 left over

carry 4

Dividing 60 by 8 gives 7, with a remainder of 4. Write the 7 on the answer line, add another 0 after the decimal place, and carry the 4.

divide 60 by 8

add another 0

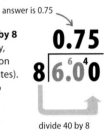

answer is 0.75

Dividing 40 by 8 gives 5 exactly, and the division ends (terminates). The answer to 6 ÷ 8 is 0.75.

divide 40 by 8

Dividing decimals

Above, short division was used to find the decimal answer for the sum 8 ÷ 6. Long division can be used to achieve the same result.

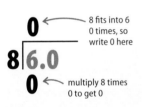

8 fits into 6 0 times, so write 0 here

multiply 8 times 0 to get 0

First, divide 8 into 6. It goes 0 times, so put a 0 above the 6. Multiply 8×0, and write the result (0) under the 6.

add decimal point

bring down a 0

divide 60 by 8

Subtract 0 from 6 to get 6, and bring down a 0. Divide 8 into 60 and put the answer, 7, after a decimal point.

multiply 8 times 7 to get 56

first remainder is 4

Work out the first remainder by multiplying 8 by 7 and subtracting this from 60. The answer is 4.

8 goes into 40 exactly 5 times

bring down a 0

divide 40 by 8

Bring down a zero to join the 4 and divide the number by 8. It goes exactly 5 times, so put a 5 above the line.

Decimals that do not end

Sometimes the answer to a division can be a decimal number that repeats without ending. This is called a "recurring" decimal. For example, here 1 is divided by 3. Both the calculations and the answers in the division become identical after the second stage, and the answer recurs endlessly.

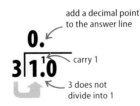

add a decimal point to the answer line

carry 1

3 does not divide into 1

3 does not divide into 1, so enter 0 on the answer line. Add a decimal point after 0, and carry 1.

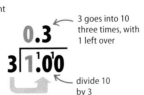

3 goes into 10 three times, with 1 left over

divide 10 by 3

10 divided by 3 gives 3, with a remainder of 1. Write the 3 on the answer line and carry the 1 to the next 0.

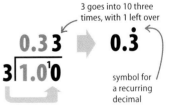

3 goes into 10 three times, with 1 left over

symbol for a recurring decimal

Dividing 10 by 3 again gives exactly the same answer as the last step. This is repeated infinitely. This type of recurring decimal is written with a dot over the recurring digit.

Fractions

A FRACTION REPRESENTS A PART OF A WHOLE NUMBER.

Fractions are a way to split up a number into equal parts.
They are written as one number over another number.

SEE ALSO

‹22–25 Division

‹38–39 Decimals in action

Ratio and
proportion   48–51 **›**

Percentages   52–53 **›**

Converting
fractions, decimals,
percentages   56–57 **›**

Writing fractions

The number on the top of a fraction
shows how many equal parts of the
whole are being dealt with, while the
number on the bottom shows the
total number of equal parts that
the whole has been divided into.

Numerator
The number of equal
parts examined.

Dividing line
This is also written as /.

Denominator
Total number of equal
parts in the whole.

Quarter
One fourth, or ¼
(a quarter), shows 1
part out of 4 equal
parts in a whole.

$\dfrac{1}{4}$

$\dfrac{1}{8}$

$\dfrac{1}{16}$

$\dfrac{1}{32}$

$\dfrac{1}{64}$ $\dfrac{1}{64}$

Eighth
⅛ (one eighth)
is 1 part out of
8 equal parts
in a whole.

Sixteenth
¹⁄₁₆ (one sixteenth)
is 1 part out of
16 equal parts
in a whole.

One thirty-second
¹⁄₃₂ (one
thirty-second)
is 1 part out of
32 equal parts
in a whole.

One sixty-fourth
¹⁄₆₄ (one
sixty-fourth) is
1 part out of 64
equal parts in
a whole.

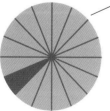

▷ **Equal parts of a whole**
The circle on the right shows how parts
of a whole can be divided in different
ways to form different fractions.

Types of fractions

A proper fraction – where the numerator is smaller than the denominator – is just one type of fraction. When the number of parts is greater than the whole, the result is a fraction that can be written in two ways – either as a "top-heavy" fraction (or improper fraction) or a "mixed" fraction.

numerator has lower value than denominator

$\dfrac{1}{4}$ ◁ **Proper fraction**
In this fraction the number of parts examined, shown on top, is less than the whole.

numerator has higher value than denominator

$\dfrac{35}{4}$ ◁ **Top-heavy fraction**
The larger numerator indicates that the parts come from more than one whole

whole number   fraction

$10\,\dfrac{1}{3}$ ◁ **Mixed fraction**
A whole number is combined with a proper fraction.

Half
½ (one half) is 1 part out of 2 equal parts in a whole.

Depicting fractions

Fractions can be illustrated in many ways, using any shape that can be divided into an equal number of parts.

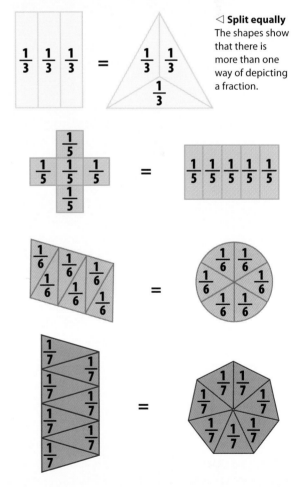

◁ **Split equally**
The shapes show that there is more than one way of depicting a fraction.

Turning top-heavy fractions into mixed fractions

A top-heavy fraction can be turned into a mixed fraction by dividing the numerator by the denominator.

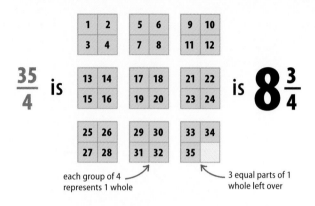

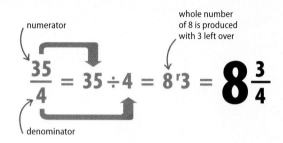

each group of 4 represents 1 whole

3 equal parts of 1 whole left over

Draw groups of four numbers – each group represents a whole. The fraction is eight whole numbers with ¾ (three quarters) left over.

numerator

denominator

whole number of 8 is produced with 3 left over

Divide the numerator by the denominator, in this case, 35 by 4.

The result is the mixed fraction 8¾ made up of the whole number 8 and 3 parts – or ¾ (three quarters) – left over.

Turning mixed fractions into top-heavy fractions

A mixed fraction can be changed into a top-heavy fraction by multiplying the whole number by the denominator and adding the result to the numerator.

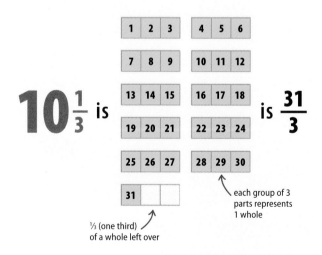

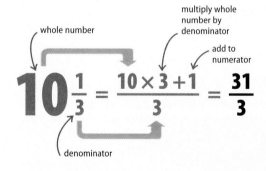

⅓ (one third) of a whole left over

each group of 3 parts represents 1 whole

Draw the fraction as ten groups of three parts with one part left over. In this way it is possible to count 31 parts in the fraction.

whole number

denominator

multiply whole number by denominator

add to numerator

Multiply the whole number by the denominator – in this case, 10 × 3 = 30. Then add the numerator.

The result is the top-heavy fraction ³¹⁄₃, with a numerator (31) greater than the denominator (3).

Equivalent fractions

The same fraction can be written in different ways. These are known as equivalent (meaning "equal") fractions, even though they look different.

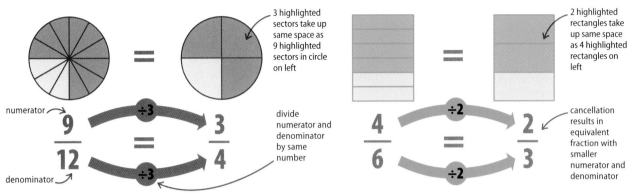

3 highlighted sectors take up same space as 9 highlighted sectors in circle on left

divide numerator and denominator by same number

2 highlighted rectangles take up same space as 4 highlighted rectangles on left

cancellation results in equivalent fraction with smaller numerator and denominator

△ **Cancellation**
Cancellation is a method used to find an equivalent fraction that is simpler than the original. To cancel a fraction divide the numerator and denominator by the same number.

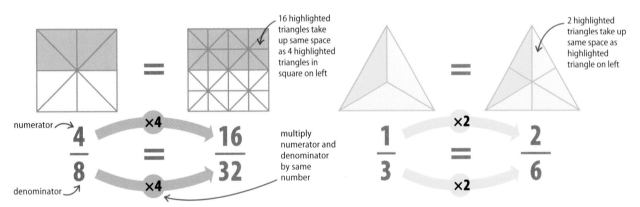

16 highlighted triangles take up same space as 4 highlighted triangles in square on left

multiply numerator and denominator by same number

2 highlighted triangles take up same space as highlighted triangle on left

△ **Reverse cancellation**
Multiplying the numerator and denominator by the same number is called reverse cancellation. This results in an equivalent fraction with a larger numerator and denominator.

Table of equivalent fractions									
$1/1 =$	$2/2$	$3/3$	$4/4$	$5/5$	$6/6$	$7/7$	$8/8$	$9/9$	$10/10$
$1/2 =$	$2/4$	$3/6$	$4/8$	$5/10$	$6/12$	$7/14$	$8/16$	$9/18$	$10/20$
$1/3 =$	$2/6$	$3/9$	$4/12$	$5/15$	$6/18$	$7/21$	$8/24$	$9/27$	$10/30$
$1/4 =$	$2/8$	$3/12$	$4/16$	$5/20$	$6/24$	$7/28$	$8/32$	$9/36$	$10/40$
$1/5 =$	$2/10$	$3/15$	$4/20$	$5/25$	$6/30$	$7/35$	$8/40$	$9/45$	$10/50$
$1/6 =$	$2/12$	$3/18$	$4/24$	$5/30$	$6/36$	$7/42$	$8/48$	$9/54$	$10/60$
$1/7 =$	$2/14$	$3/21$	$4/28$	$5/35$	$6/42$	$7/49$	$8/56$	$9/63$	$10/70$
$1/8 =$	$2/16$	$3/24$	$4/32$	$5/40$	$6/48$	$7/56$	$8/64$	$9/72$	$10/80$

Finding a common denominator

When finding the relative sizes of two or more fractions, finding a common denominator makes it much easier. A common denominator is a number that can be divided by the denominators of all of the fractions. Once this has been found, comparing fractions is just a matter of comparing their numerators.

▷ **Comparing fractions**
To work out the relative sizes of fractions, it is necessary to convert them so that they all have the same denominator. To do so, first look at the denominators of all the fractions being compared.

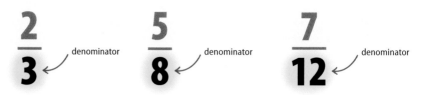

$\dfrac{2}{3}$ denominator $\dfrac{5}{8}$ denominator $\dfrac{7}{12}$ denominator

▷ **Make a list**
List the multiples – all the whole number products of each denominator – for all of the denominators. Pick a sensible stopping point for the list, such as 100.

multiples of 3   multiples of 8   multiples of 12

3, 6, 9, 12, 15, 18, 21, 24, 27, 30...   **8, 16, 24, 32, 40, 48, 56, 64, 72...**   **12, 24, 36, 48, 60, 72, 84, 96...**

▷ **Find the lowest common denominator**
List only the multiples that are common to all three sets. These numbers are called common denominators. Identify the lowest one.

lowest common denominator of 3, 8, and 12   common denominator

24, 48, 72, 96...

▷ **Convert the fractions**
Find out how many times the original denominator goes into the common denominator. Multiply the numerator by the same number. It is now possible to compare the fractions.

largest fraction   smallest fraction

$\dfrac{2}{3} \overset{\times 8}{\underset{\times 8}{=}} \dfrac{16}{24}$ $\dfrac{5}{8} \overset{\times 3}{\underset{\times 3}{=}} \dfrac{15}{24}$ $\dfrac{7}{12} \overset{\times 2}{\underset{\times 2}{=}} \dfrac{14}{24}$

original denominator goes into common denominator 8 times, so multiply both sides by 8

original denominator goes into common denominator 3 times, so multiply both sides by 3

original denominator goes into common denominator 2 times, so multiply both sides by 2

ADDING AND SUBTRACTING FRACTIONS

Just like whole numbers, it is possible to add and subtract fractions. How it is done depends on whether the denominators are the same or different.

Adding and subtracting fractions with the same denominator

To add or subtract fractions that have the same denominator, simply add or subtract their numerators to get the answer. The denominators stay the same.

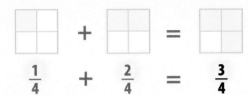

$$\frac{1}{4} + \frac{2}{4} = \frac{3}{4} \qquad\qquad \frac{7}{8} - \frac{4}{8} = \frac{3}{8}$$

To add fractions, add together only the numerators. The denominator in the result remains unchanged.

To subtract fractions, subtract the smaller numerator from the larger. The denominator in the result stays the same.

Adding fractions with different denominators

To add fractions that have different denominators, it is necessary to change one or both of the fractions so they have the same denominator. This involves finding a common denominator (see opposite).

multiply whole number by denominator then add numerator

6 is a common denominator of both 3 and 6

⁵/₆ can now be added to ²⁶/₆ as both have same denominator

remainder becomes numerator of fraction

$$4\frac{1}{3} + \frac{5}{6} \quad \frac{4\times3+1}{3} \quad \frac{13}{3} + \frac{5}{6} \quad \frac{13}{3} \stackrel{\times2}{=} \frac{26}{6} + \frac{5}{6} \quad \frac{31}{6} \Rightarrow 31\div6 = 5r1 = 5\frac{1}{6}$$

denominator stays same

denominator goes into common denominator 2 times, so multiply both sides by 2

First, turn any mixed fractions that are being added into top-heavy fractions.

▶ **It is not possible** to add fractions with different denominators, so find a common denominator.

▶ **Convert the fractions** into fractions with common denominators by multiplying.

▶ **In order to** turn the resulting top-heavy fraction back into a mixed fraction, divide the numerator by the denominator.

Subtracting fractions with different denominators

To subtract fractions with different denominators, a common denominator must be found.

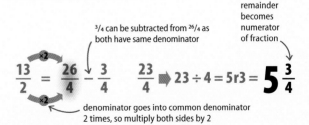

multiply whole number by denominator then add numerator

4 is a common denominator of both 2 and 4

³/₄ can be subtracted from ²⁶/₄ as both have same denominator

remainder becomes numerator of fraction

$$6\frac{1}{2} - \frac{3}{4} \quad \frac{6\times2+1}{2} \quad \frac{13}{2} - \frac{3}{4} \quad \frac{13}{2} \stackrel{\times2}{=} \frac{26}{4} - \frac{3}{4} \quad \frac{23}{4} \Rightarrow 23\div4 = 5r3 = 5\frac{3}{4}$$

denominator stays same

denominator goes into common denominator 2 times, so multiply both sides by 2

First, turn any mixed fractions in the equation into top-heavy fractions by multiplying.

▶ **The two fractions** have different denominators, so a common denominator is needed.

▶ **Convert the fractions** into fractions with common denominators by multiplying.

▶ **If necessary**, divide the numerator by the denominator to turn the top-heavy fraction back into a mixed fraction.

MULTIPLYING FRACTIONS

Fractions can be multiplied by other fractions. To multiply fractions by mixed fractions or whole numbers, they first need to be converted into top-heavy fractions.

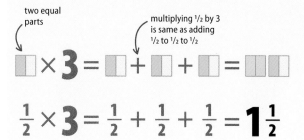

two equal parts

multiplying ½ by 3 is same as adding ½ to ½ to ½

$$\frac{1}{2} \times 3 = \frac{1}{2} + \frac{1}{2} + \frac{1}{2} = 1\frac{1}{2}$$

Imagine multiplying a fraction by a whole number as adding the fraction to itself that many times. Alternatively, imagine multiplying a whole number by a fraction as taking that portion of the whole number, here ½ of 3.

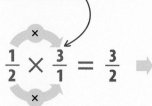

whole number a top-heavy fraction with whole number as numerator and 1 as denominator

$$\frac{1}{2} \times \frac{3}{1} = \frac{3}{2}$$

Convert the whole number to a fraction. Next, multiply both numerators together and then both denominators.

remainder becomes numerator of fraction

$$3 \div 2 = 1^r1 = 1\frac{1}{2}$$

denominator stays the same

Divide the numerator of the resulting fraction by the denominator. The answer is given as a mixed fraction.

Multiplying two proper fractions

Proper fractions can be multiplied by each other. It is useful to imagine the multiplication sign means "of" – the sum below can be expressed as "what is ½ of ¾?" and "what is ¾ of ½?".

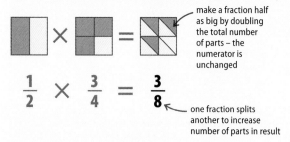

make a fraction half as big by doubling the total number of parts – the numerator is unchanged

$$\frac{1}{2} \times \frac{3}{4} = \frac{3}{8}$$

one fraction splits another to increase number of parts in result

Visually, the result of multiplying two proper fractions is that the space taken up by both together is reduced.

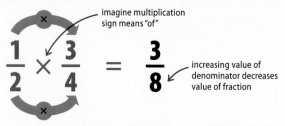

imagine multiplication sign means "of"

$$\frac{1}{2} \times \frac{3}{4} = \frac{3}{8}$$

increasing value of denominator decreases value of fraction

Multiply the numerators and the denominators. The resulting fraction answers both questions: "what is ½ of ¾?" and "what is ¾ of ½?".

Multiplying mixed fractions

To multiply a proper fraction by a mixed fraction, it is necessary to first convert the mixed fraction into a top-heavy fraction.

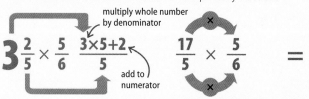

multiply whole number by denominator

$$3\frac{2}{5} \times \frac{5}{6} \quad \frac{3 \times 5 + 2}{5}$$

add to numerator

$$\frac{17}{5} \times \frac{5}{6} =$$

remainder becomes numerator of fraction

to show in its lowest form divide both numbers by 5 to get ⅚

$$\frac{85}{30} \quad 85 \div 30 = 2^r25 = 2\frac{25}{30}$$

denominator stays the same

First, turn the mixed fraction into a top-heavy fraction.

Next, multiply the numerators and denominators of both fractions to get a new fraction.

Divide the numerator of the new top-heavy fraction by its denominator. The answer is shown as a mixed fraction.

DIVIDING FRACTIONS

Fractions can be divided by whole numbers. To do so turn the whole number into a fraction, turn the fraction upside down, then multiply it by the first fraction.

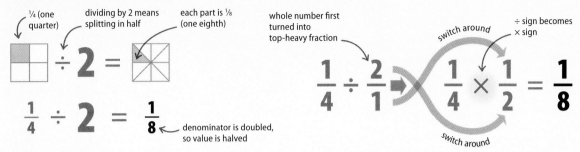

¼ (one quarter)

dividing by 2 means splitting in half

each part is ⅛ (one eighth)

whole number first turned into top-heavy fraction

switch around

÷ sign becomes × sign

$$\frac{1}{4} \div 2 =$$

$$\frac{1}{4} \div 2 = \frac{1}{8}$$

denominator is doubled, so value is halved

$$\frac{1}{4} \div \frac{2}{1} \rightarrow \frac{1}{4} \times \frac{1}{2} = \frac{1}{8}$$

switch around

Picture dividing a fraction by a whole number as splitting it into that many parts. In this example, ¼ is split in half, resulting in twice as many equal parts.

To divide a fraction by a whole number, convert the whole number into a fraction, turn the fraction upside down, and multiply both the numerators and the denominators.

Dividing two proper fractions

Proper fractions can be divided by other proper fractions by using an inverse operation. Multiplication and division are inverse operations as they are the opposite of each other.

imagine the multiplication sign means "of"

3 multiplied by ¼, or ¼ of 3, gives ¾

same as ³⁄₁

$$\frac{1}{4} \div \frac{1}{3} \text{ is same as saying } \frac{1}{4} \times 3 = \frac{3}{4}$$

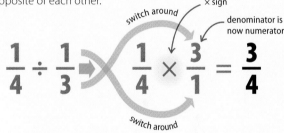

÷ sign becomes × sign

switch around

denominator is now numerator

$$\frac{1}{4} \div \frac{1}{3} \rightarrow \frac{1}{4} \times \frac{3}{1} = \frac{3}{4}$$

switch around

Dividing one fraction by another is the same as turning the second fraction upside down and then multiplying the two.

To divide two fractions use the inverse operation – turn the last fraction upside down, then multiply both the numerators and the denominators.

Dividing mixed fractions

To divide mixed fractions, first convert them into top-heavy fractions, then turn the second fraction upside down and multiply it by the first.

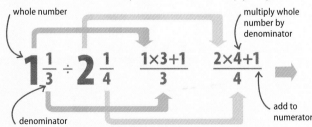

whole number

multiply whole number by denominator

$$1\frac{1}{3} \div 2\frac{1}{4} \quad \frac{1\times3+1}{3} \quad \frac{2\times4+1}{4} \rightarrow$$

denominator

add to numerator

÷ sign becomes × sign

switch around

denominator is now numerator

$$\frac{4}{3} \div \frac{9}{4} \rightarrow \frac{4}{3} \times \frac{4}{9} = \frac{16}{27}$$

switch around

First, turn both the mixed fractions into top-heavy fractions by multiplying the whole number by the denominator and adding the numerator.

Divide the two fractions by turning the second fraction upside down, then multiplying both the numerators and the denominators.

Ratio and proportion

SEE ALSO
❰ **18–21** Multiplication
❰ **22–25** Division
❰ **40–47** Fractions

RATIO COMPARES THE SIZE OF TWO QUANTITIES. PROPORTION COMPARES THE RELATIONSHIP BETWEEN TWO SETS OF QUANTITIES.

Ratios show how much bigger one thing is than another. Two things are in proportion when a change in one causes a related change in the other.

Writing ratios

Ratios are written as two or more numbers with a colon between each. For example, a fruit bowl in which the ratio of apples to pears is 2 : 1 means that there are 2 apples for every 1 pear in the bowl.

◁ **Supporters**
This group represents fans of two football clubs, the "greens" and the "blues".

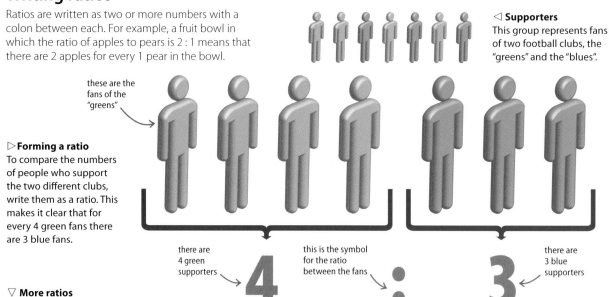

these are the fans of the "greens"

▷ **Forming a ratio**
To compare the numbers of people who support the two different clubs, write them as a ratio. This makes it clear that for every 4 green fans there are 3 blue fans.

there are 4 green supporters

this is the symbol for the ratio between the fans

there are 3 blue supporters

▽ **More ratios**
The same process applies to any set of data that needs comparing. Here are more groups of fans, and the ratios they represent.

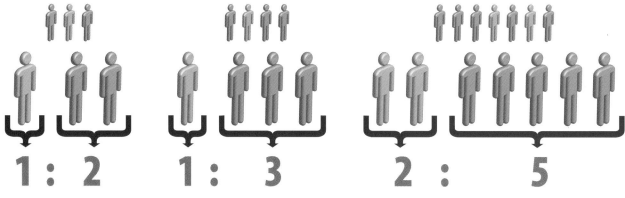

△ **1 : 2**
One fan of the greens and 2 fans of the blues can be compared as the ratio 1 : 2. This means that in this case there are twice as many fans of the blues as of the greens.

△ **1 : 3**
One fan of the greens and 3 fans of the blues can be shown as the ratio 1 : 3, which means that, in this case, there are three times more blue fans than green fans.

△ **2 : 5**
Two fans of the greens and 5 fans of the blues can be compared as the ratio 2 : 5. There are more than twice as many fans of the blues as of the greens.

Finding a ratio

Large numbers can also be written as ratios. For example, to find the ratio between 1 hour and 20 minutes, convert them into the same unit, then cancel these numbers by finding the highest number that divides into both.

minutes are the smaller unit

1 hour is the same as 60 minutes, so convert

this is the symbol for ratio

ratios show information in the same way as fractions do

20 minutes is $\frac{1}{3}$ of an hour

20 mins, 60 mins   20 : 60   1 : 3

$60 \div 20 = 3$
$20 \div 20 = 1$

Convert one of the quantities so that both have the same units. In this example use minutes.

Write as a ratio by inserting a colon between the two quantities.

Cancel the units to their lowest terms. Here both sides divide exactly by 20 to give the ratio 1 : 3.

Working with ratios

Ratios can represent real values. In a scale, the small number of the ratio is the value on the scale model, and the larger is the real value it represents.

▷ **Scaling down**
1 : 50,000cm is used as the scale on a map. Find out what a distance of 1.5cm represents on this map.

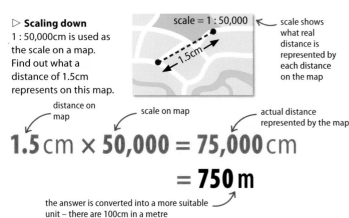

scale = 1 : 50,000

scale shows what real distance is represented by each distance on the map

▷ **Scaling up**
The plan of a microchip has the scale 40 : 1. The length of the plan is 18cm. The scale can be used to find the length of the actual microchip.

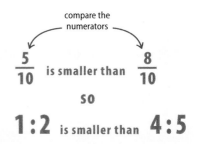

distance on map

scale on map

actual distance represented by the map

$1.5\,\text{cm} \times 50,000 = 75,000\,\text{cm}$
$= 750\,\text{m}$

the answer is converted into a more suitable unit – there are 100cm in a metre

length of plan

divide by scale to find actual size

actual length of microchip

$18\,\text{cm} \div 40 = 0.45\,\text{cm}$

Comparing ratios

Converting ratios into fractions allows their size to be compared. To compare the ratios 4 : 5 and 1 : 2, write them as fractions with the same denominator.

compare the numerators

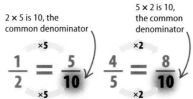

$$1 : 2 = \frac{1}{2}$$
fraction that represents ratio 1 : 2

and

$$4 : 5 = \frac{4}{5}$$
fraction that represents ratio 4 : 5

2 × 5 is 10, the common denominator

5 × 2 is 10, the common denominator

$$\frac{1}{2} \overset{\times 5}{\underset{\times 5}{=}} \frac{5}{10} \qquad \frac{4}{5} \overset{\times 2}{\underset{\times 2}{=}} \frac{8}{10}$$

$$\frac{5}{10} \text{ is smaller than } \frac{8}{10}$$

so

1 : 2 is smaller than **4 : 5**

First write each ratio as a fraction, placing the smaller quantity in each above the larger quantity.

Convert the fractions so that they both have the same denominator, by multiplying the first fraction by 5 and the second by 2.

As the fractions now share a denominator, their sizes can be compared, making it clear which ratio is bigger.

PROPORTION

Two quantities are in proportion when a change in one causes a change in the other. Two examples of this are direct and indirect (also called inverse) proportion.

Direct proportion

Two quantities are in direct proportion if the ratio between them is always the same. This means, for example, that if one quantity doubles then so does the other.

▷ **Direct proportion**
This table and graph show the directly proportional relationship between the number of gardeners and the number of trees planted.

Gardeners	Trees
1	2
2	4
3	6

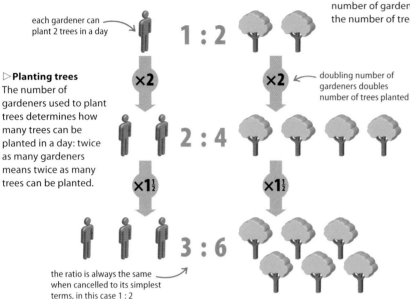

each gardener can plant 2 trees in a day

1 : 2

×2   ×2 ← doubling number of gardeners doubles number of trees planted

▷**Planting trees**
The number of gardeners used to plant trees determines how many trees can be planted in a day: twice as many gardeners means twice as many trees can be planted.

2 : 4

×1½   ×1½

3 : 6

the ratio is always the same when cancelled to its simplest terms, in this case 1 : 2

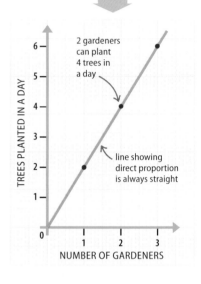

2 gardeners can plant 4 trees in a day

line showing direct proportion is always straight

TREES PLANTED IN A DAY

NUMBER OF GARDENERS

Indirect proportion

Two quantities are in indirect proportion if their product (the answer when they are multiplied by each other) is always the same. So if one quantity doubles, the other quantity halves.

▷ **Indirect proportion**
This table and graph show the indirectly proportional relationship between the vans used and the time taken to deliver the parcels.

Vans	Days
1	8
2	4
4	2

1 van takes 8 days to deliver some parcels

1 : 8

×2   ÷2

▷ **Delivering parcels**
The number of vans used to deliver parcels determines how many days it takes to deliver the parcels. Twice as many vans means half as many days to deliver.

2 : 4

2 vans take 4 days to deliver the parcels

if the number of vans doubles then it takes half the time to deliver the parcels

×2   ÷2

4 : 2

the product of the number of vans and days is always the same: 8

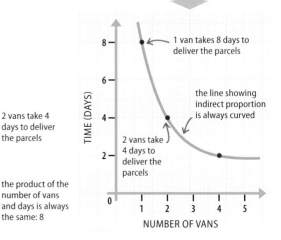

1 van takes 8 days to deliver the parcels

the line showing indirect proportion is always curved

2 vans take 4 days to deliver the parcels

TIME (DAYS)

NUMBER OF VANS

Dividing in a given ratio

A quantity can be divided into two, three, or more parts, according to a given ratio. This example shows how to divide 20 people into the ratios 2 : 3 and 6 : 3 : 1.

DIVIDING INTO A
TWO-PART RATIO

2 : 3

These are the ratios to divide the people into.

DIVIDING INTO A
THREE-PART RATIO

6 : 3 : 1

total number of parts in the ratio

$2 + 3 = 5$

Add the different parts of the ratio to find the total parts.

$6 + 3 + 1 = 10$

number of parts in the ratio

total number of people

$20 \div 5 = 4$

Divide the number of people by the parts of the ratio.

total number of people — total number of parts in the ratio

$20 \div 10 = 2$

2 in the ratio

$2 \times 4 = 8$

3 in the ratio

$3 \times 4 = 12$

12 people represented by 3 in the ratio

8 people represented by 2 in the ratio

Multiply each part of the ratio by this quantity to find the size of the groups the ratios represent.

6 in the ratio

$6 \times 2 = 12$

3 in the ratio

$3 \times 2 = 6$

1 in the ratio

$1 \times 2 = 2$

12 people represented by 6 in the ratio

6 people represented by 3 in the ratio

2 people represented by 1 in the ratio

Proportional quantities

Proportion can be used to solve problems involving unknown quantities. For example, if 3 bags contain 18 apples, how many apples do 5 bags contain?

total number of apples — bags — apples per bag

$18 \div 3 = 6$

apples per bag — number of bags

$6 \times 5 = 30$ ← total

There is a total of 18 apples in 3 bags. Each bag contains the same number of apples.

To find out how many apples there are in 1 bag, divide the total number of apples by the number of bags.

To find the number of apples in 5 bags, multiply the number of apples in 1 bag by 5.

% Percentages

A PERCENTAGE SHOWS AN AMOUNT AS A PART OF 100.

SEE ALSO

❮ **38–39** Decimals in action

❮ **40–47** Fractions

Ratio and proportion **48–51** ❯

Rounding off **62–63** ❯

Any number can be written as a part of 100 or a percentage. Per cent means "per hundred", and it is a useful way of comparing two or more quantities. The symbol "%" is used to indicate a percentage.

Parts of 100

The simplest way to start looking at percentages is by dealing with a block of 100 units, as shown in the main image. These 100 units represent the total number of people in a school. This total can be divided into different groups according to the proportion of the total 100 they represent.

100%

▷ **This is simply** another way of saying "everybody" or "everything". Here, all 100 figures – 100% – are blue.

50%

▷ **This group** is equally divided between 50 blue and 50 purple figures. Each represents 50 out of 100 or 50% of the total. This is the same as half.

1%

▷ **In this group** there is only 1 blue figure out of 100, or 1%.

FEMALE TEACHERS **10%** or 10 out of 100

MALE STUDENTS **19%** or 19 out of 100

MALE TEACHERS **5%** or 5 out of 100

△ **Adding up to 100**
Percentages are an effective way to show the component parts of a total. For example, male teachers (blue) account for 5% (5 out of 100) of the total.

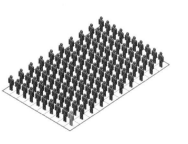

FEMALE STUDENTS
66% or
66 out of 100

WORKING WITH PERCENTAGES

A percentage is simply a part of a whole, expressed as a part of 100. There are two main ways of working with percentages: the first is finding a percentage of a given amount, and the second is finding what percentage one number is of another number.

Calculating percentages

This example shows how to find the percentage of a quantity, in this case 25% of a group of 24 people.

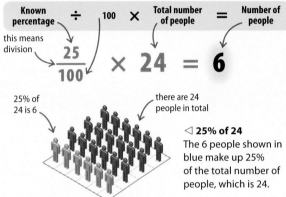

| Known percentage | ÷ | 100 | × | Total number of people | = | Number of people |

this means division

$$\frac{25}{100} \times 24 = 6$$

25% of 24 is 6

there are 24 people in total

◁ **25% of 24**
The 6 people shown in blue make up 25% of the total number of people, which is 24.

This example shows how to find what percentage one number is of another number, in this case 48 people out of a group of 112 people.

| Number of people | ÷ | Total number of people | × | 100 | = | Percentage of total number |

$$\frac{48}{112} \times 100 = 42.86$$

answer is rounded to 2 decimal places

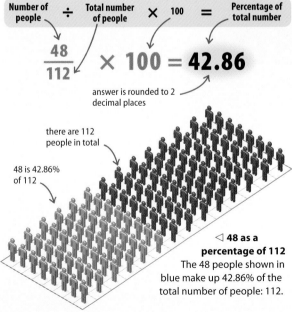

there are 112 people in total

48 is 42.86% of 112

◁ **48 as a percentage of 112**
The 48 people shown in blue make up 42.86% of the total number of people: 112.

▽ **Examples of percentages**
Percentages are a simple and accessible way to present information, which is why they are often used by the media.

Percentage	Facts
97%	of the world's animals are invertebrates
92.5%	of an Olympic gold medal is composed of silver
70%	of the world's surface is covered in water
66%	of the human body is water
61%	of the world's oil is in the Middle East
50%	of the world's population live in cities
21%	of the air is oxygen
6%	of the world's land surface is covered in rainforest

PERCENTAGES AND QUANTITIES

Percentages are a useful way of expressing a value as a proportion of the total number. If two out of three of a percentage, value, and total number are known, it is possible to find out the missing quantity using arithmetic.

Finding an amount as a % of another amount

Out of 12 pupils in a class, 9 play a musical instrument. To find the known value (9) as a percentage of the total (12), divide the known value by the total number and multiply by 100.

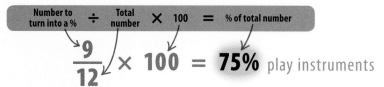

$$\frac{9}{12} \times 100 = 75\% \text{ play instruments}$$

Divide the known number by the total number (9 ÷ 12 = 0.75).

Multiply the result by 100 to get the percentage (0.75 × 100 = 75).

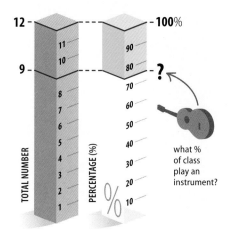

what % of class play an instrument?

Finding the total number from a %

In a class, 7 children make up 35% of the total. To find the total number of students in the class, divide the known value (7) by the known percentage (35) and multiply by 100.

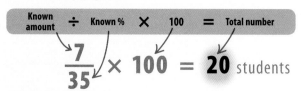

$$\frac{7}{35} \times 100 = 20 \text{ students}$$

Divide the known amount by the known percentage (7 ÷ 35 = 0.2).

Multiply the result by 100 to get the total number (0.2 × 100 = 20).

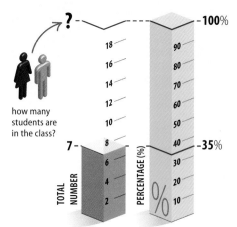

how many students are in the class?

Percentages

Percentages are all around us – in shops, in newspapers, on TV – everywhere. Many things in everyday life are measured and compared in percentages – how much an item is reduced in a sale; what the interest rate is on a mortgage or a bank loan; or how efficient a light bulb is by the percentage of electricity it converts to light. Percentages are even used to show how much of the recommended daily intake of vitamins and other nutrients is in food products.

SALE

25% OFF

PERCENTAGE CHANGE

If a value changes by a certain percentage, it is possible to calculate the new value. Conversely, when a value changes by a known amount, it is possible to work out the percent increase or decrease compared to the original.

Finding a new value from a % increase or decrease

To find how a 55% increase or decrease affects the value of 40, first work out 55% of 40. Then add to or subtract from the original to get the new value.

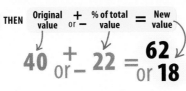

$$\frac{55}{100} \times 40 = 22 \quad \textbf{THEN} \quad 40 \begin{array}{c} + \\ \text{or} \\ - \end{array} 22 = \begin{array}{c} \textbf{62} \\ \text{or} \\ \textbf{18} \end{array}$$

Known % ÷ 100 × Original value = % of total value

Divide the known % by 100 (55 ÷ 100 = 0.55).

Multiply the result by the original value (0.55 × 40 = 22).

Original value + or − % of total value = New value

Add the original value to 22 to find the % increase, or **subtract** 22 to find the % decrease.

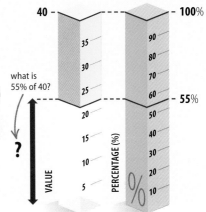

what is 55% of 40?

Finding an increase in a value as a %

The price of a doughnut in the school canteen has risen 30p – from 99p last year to £1.29 this year. To find the increase as a percent, divide the increase in value (30) by the original value (99) and multiply by 100.

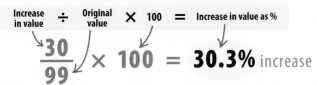

Increase in value ÷ Original value × 100 = Increase in value as %

$$\frac{30}{99} \times 100 = \textbf{30.3\%} \ \text{increase}$$

Divide the increase in value by the original value (30 ÷ 99 = 0.303).

Multiply the result by 100 to find the percentage (0.303 × 100 = 30.3), and round to 3 significant figures.

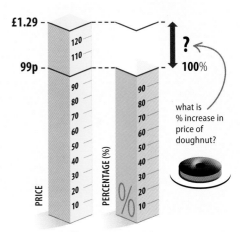

what is % increase in price of doughnut?

Finding a decrease in a value as a %

There was an audience of 245 at the school play last year, but this year only 209 attended – a decrease of 36. To find the decrease as a percent, divide the decrease in value (36) by the original value (245) and multiply by 100.

Decrease in value ÷ Original value × 100 = Decrease in value as %

$$\frac{36}{245} \times 100 = \textbf{14.7\%} \ \text{decrease}$$

Divide the decrease in value by the original value (36 ÷ 245 = 0.147).

Multiply the result by 100 to find the percentage (0.147 × 100 = 14.7), and round to 3 significant figures.

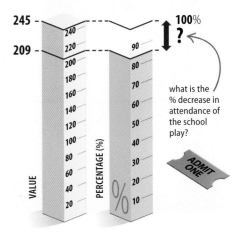

what is the % decrease in attendance of the school play?

↻ Converting fractions, decimals, and percentages

SEE ALSO
❰ 38–39 Decimals in action
❰ 40–47 Fractions
❰ 52–55 Percentages

DECIMALS, FRACTIONS, AND PERCENTAGES ARE DIFFERENT WAYS OF WRITING THE SAME NUMBER.

The same but different

Sometimes a number shown one way can be shown more clearly in another way. For example, if 20% is the mark required to pass an exam, this is the same as saying that 1/5 of the answers in an exam need to be answered correctly to achieve a pass mark or that the minimum score for a pass is 0.2 of the total.

75%

PERCENTAGE

A percentage shows a number as a proportion of 100.

Changing a **decimal** into a **percentage**

To change a decimal into a percentage, multiply by 100.

$$0.75 \rightarrow 75\%$$

$$0.75 \times 100 = 75\%$$

Decimal   **Multiply by 100**   **Percentage**

decimal point in 0.75 moved two places to right to make 75

▷ **All change**
The three ways of writing the same number are shown here: decimal (0.75), fraction (¾), and percentage (75%). They look different, but they all represent the same proportion of an amount.

Changing a **percentage** into a **decimal**

To change a percentage into a decimal, divide it by 100.

$$75\% \rightarrow 0.75$$

decimal point added two places to left of last digit

$$75\% \div 100 = 0.75$$

Percentage   **Divide by 100**   **Decimal**

Changing a **percentage** into a **fraction**

To change a percentage into a fraction, write it as a fraction of 100 and then cancel it down to simplify it, if possible.

$$75\% \rightarrow \frac{3}{4}$$

divide by highest number that goes into 75 and 100

$$75\% \rightarrow \frac{75}{100}$$

÷25
÷25

$$\frac{3}{4}$$

Percentage

Turn the percentage into the numerator of a fraction with 100 as the denominator.

Fraction cancelled down into its lowest terms.

0.75

DECIMAL

A decimal is simply a number that is not whole. It always contains a decimal point.

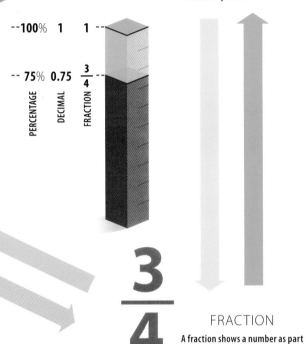

--100% **1**   **1**--

-- 75% **0.75** $\frac{3}{4}$--

PERCENTAGE   DECIMAL   FRACTION

$\frac{3}{4}$

FRACTION

A fraction shows a number as part of an equally divided whole.

Everyday numbers to remember

Many decimals, fractions, and percentages are used in everyday life – some of the more common ones are shown here.

Decimal	Fraction	%	Decimal	Fraction	%
0.1	$\frac{1}{10}$	10%	0.625	$\frac{5}{8}$	62.5%
0.125	$\frac{1}{8}$	12.5%	0.666	$\frac{2}{3}$	66.7%
0.25	$\frac{1}{4}$	25%	0.7	$\frac{7}{10}$	70%
0.333	$\frac{1}{3}$	33.3%	0.75	$\frac{3}{4}$	75%
0.4	$\frac{2}{5}$	40%	0.8	$\frac{4}{5}$	80%
0.5	$\frac{1}{2}$	50%	1	$\frac{1}{1}$	100%

Changing a **decimal** into a **fraction**

First, make the fraction's denominator (its bottom part) 10, 100, 1,000, and so on for every digit after the decimal point.

$$0.75 \implies \frac{3}{4}$$

divide by highest number that goes into 75 and 100

$$0.75 \implies \frac{75}{100} \quad \overset{\div 25}{\underset{\div 25}{\implies}} \quad \frac{3}{4}$$

Decimal number with two digits after the decimal point.

Count the decimal places – if there is 1 digit, the denominator is 10; if there are 2, it is 100. The numerator is the number after the decimal point.

Cancel the fraction down into its lowest possible terms.

Changing a **fraction** into a **percentage**

To change a fraction into a percentage, change it to a decimal and then multiply it by 100.

$$\frac{3}{4} \implies \mathbf{75\%}$$

divide the denominator (4) into the numerator (3)

$$\frac{3}{4} \implies 3 \div 4 = 0.75 \implies 0.75 \times 100 = \mathbf{75\%}$$

Fraction

Divide the numerator by the denominator.

Multiply by 100

Changing a **fraction** into a **decimal**

Divide the fraction's denominator (its bottom part) into the fraction's numerator (its top part).

$$\frac{3}{4} \implies \mathbf{0.75}$$

numerator   denominator

$$\frac{3}{4} = 3 \div 4 = \mathbf{0.75}$$

Fraction

Divide the numerator by the denominator.

Decimal

Mental maths

EVERYDAY PROBLEMS CAN BE SIMPLIFIED SO THAT THEY CAN
BE EASILY DONE WITHOUT USING A CALCULATOR.

SEE ALSO
‹ **18–21** Multiplication
‹ **22–25** Division
Using
a calculator   **64–65** ›

MULTIPLICATION

Multiplying by some numbers can be easy. For example, to multiply by 10 either
add a 0 or move the decimal point one place to the right. Also, to multiply by 20,
again multiply by 10 and then double the answer.

▷ **Multiplying by 10**
A sports club hired 2
people last year, but
this year it needs to
hire 10 times that
amount. How many
staff members will it
recruit this year?

2 staff members
recruited last year

20 new members
of staff

2   ×10   **20**

number of staff members
recruited last year

2 × 10

zero added to give 20, which is
new number of staff members

◁ **Finding the
answer**
To multiply 2 by 10
add a 0 to the 2.
Multiplying 2
people by 10 results
in an answer of 20.

▷ **Multiplying by 20**
A shop is selling
t-shirts for the price
of $/£1.20 each. How
much will the price
be for 20 t-shirts?

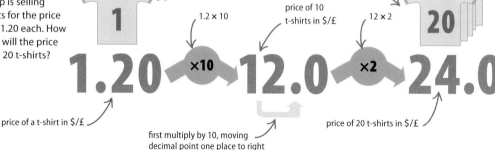

t-shirt for sale

20 t-shirts for sale

price of 10
t-shirts in $/£

1   1.2 × 10   12 × 2   **20**

1.20   ×10   **12.0**   ×2   **24.0**

price of a t-shirt in $/£

first multiply by 10, moving
decimal point one place to right

price of 20 t-shirts in $/£

◁ **Finding the
answer**
First multiply the
price by 10, here
by moving the
decimal point one
place to the right,
and then double
that to give the
final price of $/£24.

▷ **Multiplying by 25**
An athlete runs 16km
a day. If the athlete
runs the same
distance every day for
25 days, how far will
he run in total?

athlete runs every day

athlete runs
every day for 25 days

16 × 100

1,600km run
in 100 days

1,600 ÷ 4

16   ×100   **1,600**   ÷4   **400**

16km run in a day

400km run in 25 days

◁ **Finding the
answer**
First multiply the
16km for one day by
100, to give 1,600km
for 100 days, then
divide by 4 to give
the answer over
25 days.

▽ Multiplication using decimals

Decimals appear to complicate the problem, but they can be ignored until the final stage. Here the amount of carpet required to cover a floor needs to be calculated.

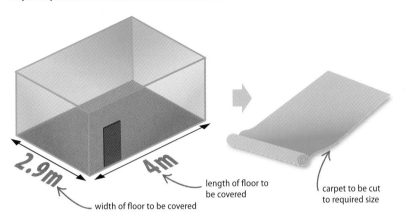

2.9m — width of floor to be covered

4m — length of floor to be covered

carpet to be cut to required size

Checking the answer

As 2.9 is almost 3, multiplying 3×4 is a good way to check that the calculation to 2.9×4 is correct.

symbol for approximately equal to

$$2.9 \approx 3 \text{ and}$$
$$3 \times 4 = 12$$

close to real answer of 11.6

$$\text{so } 2.9 \times 4 \approx 12$$

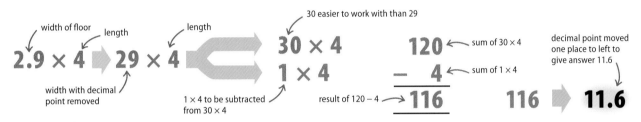

width of floor — length

$$2.9 \times 4 \Rightarrow 29 \times 4$$

width with decimal point removed

length

$$30 \times 4$$
$$1 \times 4$$

30 easier to work with than 29

1×4 to be subtracted from 30×4

$$120 \leftarrow \text{sum of } 30 \times 4$$
$$- \quad 4 \leftarrow \text{sum of } 1 \times 4$$
$$116$$

result of $120 - 4 \rightarrow$

$$116 \Rightarrow 11.6$$

decimal point moved one place to left to give answer 11.6

First, take away the decimal point from the 2.9 to make the calculation 29×4.

Change 29×4 to 30×4, as it is easier to work out. Write 1×4 below as it is the difference between 29×4 and 30×4.

Subtract 4 (product of 1×4) from 120 (product of 30×4) to give the answer of 116 (product of 29×4).

Move the decimal point one place to the left as it was moved one place to the right in the first step.

Top tricks

The multiplication tables of several numbers reveal patterns of multiplications. Here are two good mental tricks to remember when multiplying the 9 and 11 times tables.

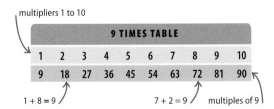

multipliers 1 to 10

9 TIMES TABLE									
1	2	3	4	5	6	7	8	9	10
9	18	27	36	45	54	63	72	81	90

$1 + 8 = 9$ $7 + 2 = 9$ multiples of 9

△ Two digits are added together

The two digits that make up the first 10 multiples of 9 each add up to 9. The first digit of the multiple (such as 1, in 18) is always 1 less than the multiplier (2).

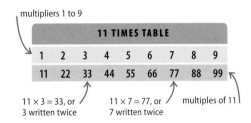

multipliers 1 to 9

11 TIMES TABLE								
1	2	3	4	5	6	7	8	9
11	22	33	44	55	66	77	88	99

$11 \times 3 = 33$, or 3 written twice $11 \times 7 = 77$, or 7 written twice   multiples of 11

△ Digit is written twice

To multiply by 11, merely repeat the two multipliers together. For example, 4×11 is two 4s or 44. It works all the way up to $9 \times 11 = 99$, which is 9 written twice.

DIVISION

Dividing by 10 or 5 is straightforward. To divide by 10, either delete a 0 or move the decimal point one place to the left. To divide by 5, again divide by 10 and then double the answer. Using these rules, work out the divisions in the following two examples.

▷ **Dividing by 10**
In this example, 160 travel vouchers are needed to hire a 10-seat mini bus. How many travel vouchers are needed for each of the 10 children to travel on the bus?

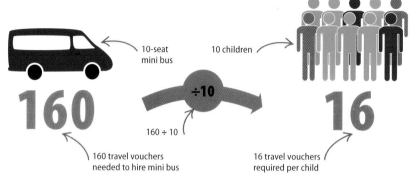

10-seat mini bus

10 children

÷10

160 ÷ 10

160 travel vouchers needed to hire mini bus

16 travel vouchers required per child

◁ **How many each?**
To find the number of travel vouchers for each child, divide the total of 160 by 10 by deleting a 0 from the 160. It gives the answer of 16 travel vouchers each.

▷ **Dividing by 5**
The cost of admission to a zoo for a group of five children is 75 tokens. How many tokens are needed for 1 of the 5 children to enter the zoo?

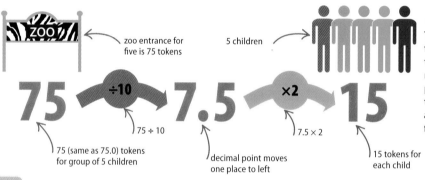

zoo entrance for five is 75 tokens

5 children

÷10

75 ÷ 10

75 (same as 75.0) tokens for group of 5 children

×2

7.5 × 2

decimal point moves one place to left

15 tokens for each child

◁ **How many each?**
To find the admission for 1 child, divide the total of 75 by 10 (by moving a decimal point in 75 one place to the left) to give 7.5, and then double that for the answer of 15.

LOOKING CLOSER

Top tips

There are various mental tricks to help with dividing larger or more complicated numbers. In the three examples below, there are tips on how to check whether very large numbers can be divided by 3, 4, and 9.

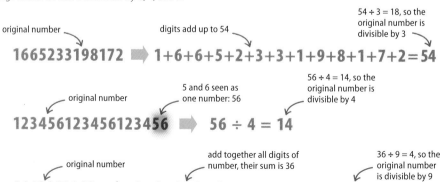

▷ **Divisible by 3**
Add together all of the digits in the number. If the total is divisible by 3, the original number is too.

original number

digits add up to 54

1665233198172 ⟹ 1+6+6+5+2+3+3+1+9+8+1+7+2 = 54

$54 ÷ 3 = 18$, so the original number is divisible by 3

▷ **Divisible by 4**
If the last two digits are taken as one single number, and it is divisible by 4, the original number is too.

original number

5 and 6 seen as one number: 56

123456123456123456 ⟹ 56 ÷ 4 = 14

$56 ÷ 4 = 14$, so the original number is divisible by 4

▷ **Divisible by 9**
Add together all of the digits in the number. If the total is divisible by 9, the original number is too.

original number

add together all digits of number, their sum is 36

1643951142 ⟹ 1+6+4+3+9+5+1+1+4+2 = 36

$36 ÷ 9 = 4$, so the original number is divisible by 9

PERCENTAGES

A useful method of simplifying calculations involving percentages is to reduce one difficult percentage into smaller and easier-to-calculate parts. In the example below, the smaller percentages include 10% and 5%, which are easy to work out.

▷ **Adding 17.5 per cent**
Here a shop wants to charge $/£480 for a new bike. However, the owner of the shop has to add a sales tax of 17.5 per cent to the price. How much will it then cost?

bike for sale before sales tax

bike for sale after sales tax

+17.5%

sales tax

480

original price in $/£

final price in $/£

564

sales tax

17.5% of 480

original price of bike

2.5% of 480 is half of 5% of 480, which is half of 10% of 480

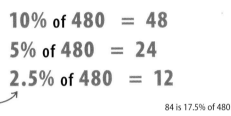

10% of 480 = 48

5% of 480 = 24

2.5% of 480 = 12

48

24

+ 12

results added together

84 is 17.5% of 480 → **84**

First, write down the percentage price increase required and the original price of the bike.

▷ **Next,** reduce 17.5% into the easier stages of 10%, 5%, and 2.5% of £/$480, and calculate their values.

▷ **The sum** of 48, 24, and 12 is 84, so $/£84 is added to $/£480 for a price of $/£564.

Switching

A percentage and an amount can both be "switched", to produce the same result with each switch. For example, 50% of 10, which is 5, is exactly the same as 10% of 50, which is 5 again.

Progression

A progression involves dividing the percentage by a number and then multiplying the amount by the same number. For example, 40% of 10 is 4. Dividing this 40% by 2 and multiplying 10 by 2 results in 20% of 20, which is also 4.

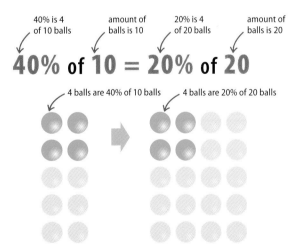

20% is 2 of 10 balls

amount of balls is 10

10% is 2 of 20 balls

amount of balls is 20

20% of 10 = 10% of 20

2 balls are 20% of 10 balls

2 balls are 10% of 20 balls

40% is 4 of 10 balls

amount of balls is 10

20% is 4 of 20 balls

amount of balls is 20

40% of 10 = 20% of 20

4 balls are 40% of 10 balls

4 balls are 20% of 20 balls

 Rounding off

THE PROCESS OF ROUNDING OFF INVOLVES REPLACING ONE
NUMBER WITH ANOTHER TO MAKE IT MORE PRACTICAL TO USE.

SEE ALSO
❮ **38–39** Decimals in action
❮ **58–59** Mental mathematics

Estimation and approximation

In many practical situations, an exact answer is not needed, and it is easier
to find an estimate based on rounding off (approximation). The general
principle of rounding off is that a number at or above the midpoint of a
group of numbers, such as the numbers 15-19 in the group 10-20, rounds
up, while a number below the midpoint rounds down.

▽ **Rounding to the nearest 10**
The midpoint between any two 10s is 5. If
the last digit of each number is 5 or over it
rounds up, otherwise it rounds down.

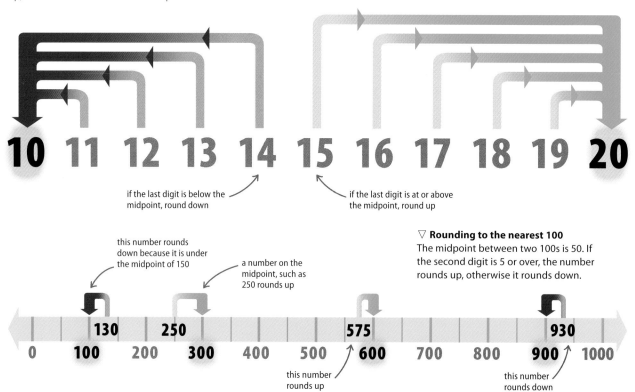

if the last digit is below the
midpoint, round down

if the last digit is at or above
the midpoint, round up

this number rounds
down because it is under
the midpoint of 150

a number on the
midpoint, such as
250 rounds up

▽ **Rounding to the nearest 100**
The midpoint between two 100s is 50. If
the second digit is 5 or over, the number
rounds up, otherwise it rounds down.

this number
rounds up

this number
rounds down

LOOKING CLOSER

Approximately equal

Many measurements are given as
approximations, and numbers are sometimes
rounded to make them easier to use. An
"approximately equals" sign is used to show when
numbers have been rounded up or down. It looks
similar to a normal equals sign (=) but with
curved instead of straight lines.

wavy lines mean
"approximately"

$$31 \approx 30 \quad \text{and} \quad 187 \approx 200$$

△ **Approximately equal to**
The "approximately equals" sign shows that the two
sides of the sign are approximately equal instead of
equal. So 31 is approximately equal to 30, and 187 is
approximately equal to 200.

Decimal places

Any number can be rounded to the appropriate number of decimal places. The choice of how many decimal places depends on what the number is used for and how exact an end result is required.

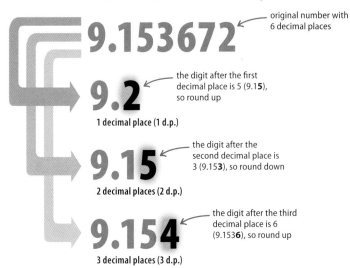

original number with 6 decimal places

9.153672

9.2 the digit after the first decimal place is 5 (9.1**5**), so round up

1 decimal place (1 d.p.)

9.15 the digit after the second decimal place is 3 (9.15**3**), so round down

2 decimal places (2 d.p.)

9.154 the digit after the third decimal place is 6 (9.153**6**), so round up

3 decimal places (3 d.p.)

How many decimal places?

The more decimal places, the more accurate the number. This table shows the accuracy that different numbers of decimal places represent. For example, a distance in kilometres to 3 decimal places would be accurate to a thousandth of a kilometre, which is equal to a metre.

Decimal places	Rounded to	Example
1	$^1/_{10}$	1.1km
2	$^1/_{100}$	1.14km
3	$^1/_{1000}$	1.135km

Significant figures

A significant figure in a number is a digit that counts. The digits 1 to 9 are always significant, but 0 is not. However, 0 becomes significant when it occurs between two significant figures, or if an exact answer is needed.

1 significant figure
200
Real value anywhere between 150–249

2 significant figures
200
Real value anywhere between 195–204

3 significant figures
200
Real value anywhere between 199.5–200.4

◁ **Significant zeros**
The answer 200 could be the result of rounding to 1, 2, or 3 significant figures (s.f.). Below each example is the range in which its true value lies.

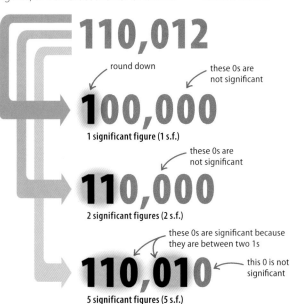

110,012

round down

these 0s are not significant

100,000
1 significant figure (1 s.f.)

these 0s are not significant

110,000
2 significant figures (2 s.f.)

these 0s are significant because they are between two 1s

this 0 is not significant

110,010
5 significant figures (5 s.f.)

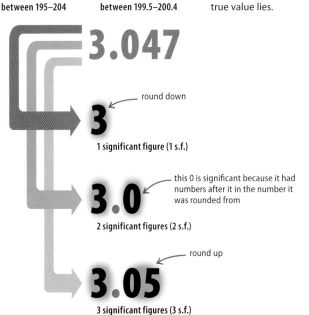

3.047

round down

3
1 significant figure (1 s.f.)

this 0 is significant because it had numbers after it in the number it was rounded from

3.0
2 significant figures (2 s.f.)

round up

3.05
3 significant figures (3 s.f.)

Using a Calculator

CALCULATORS ARE MACHINES THAT WORK OUT THE ANSWERS
TO SOME MATHEMATICAL PROBLEMS.

SEE ALSO

Tools in geometry **74–75 ›**

Collecting and
organizing data **196–197 ›**

**Calculators are designed to make maths easier, but there are
a few things to be aware of when using them.**

Introducing the calculator

A modern calculator is a handheld electronic device
that is used to find the answers to mathematical
problems. Most calculators are operated in a similar
way (as described here), but it may be necessary to
read the instructions for a particular model.

Using a calculator

Be careful that functions are entered in the correct order, or
the answers the calculator gives will be wrong.

For example, to find the answer to the sum:

$$(7 + 2) \times 9 =$$

Enter these keys, making sure to include all parts of
the sum, including the brackets.

 = **81**

Not

 = **25** ← calculator does
sum 2 × 9 = 18,
then 18 + 7 = 25

Estimating answers

Calculators can only give answers according to the keys that
have been pressed. It is useful to have an idea of what answer
to expect as a small mistake can give a very wrong answer.

For example

2 0 0 6 × 1 9 8

must be close to   this would give the
answer 400,000

2 0 0 0 × 2 0 0

So if the calculator gives the answer **40,788** it is clear that
the sum has not been entered correctly – the sum entered
had one "0" missing from what was intended:

FREQUENTLY USED KEYS

ON
This button turns the calculator on – most
calculators turn themselves off automatically if they are left
unused for a certain period of time.

Number pad
This contains the basic numbers that are needed for
maths. These buttons can be used individually or in groups to
create larger numbers.

Standard arithmetic keys
These cover all the basic mathematical functions:
multiplication, division, addition, and subtraction, as well as
the essential equals sign.

Decimal point
This key works in the same way as a written decimal
point – it separates whole numbers from decimals. It is entered
in the same way as any of the number keys.

Cancel
The cancel key clears all recent entries from the
memory. This is useful when starting a new calculation, as it
makes sure no unwanted values are retained.

Delete
This clears the last value that was entered into the
calculator, rather than wiping everything from the memory. It
is sometimes labelled "CE" (clear entry).

Recall button
This recalls a value from the calculator's memory
– it is useful for sums with many parts that use numbers or
stages from earlier in the working out.

FUNCTION KEYS

Cube
This is a short cut to cubing a number, without having to key in a number multiplied by itself, and then multiplied by itself again. Key in the number to be cubed, then press this button.

ANS
Pressing this key gives the answer to the last sum that was entered. It is useful for sums with many steps.

Square root
This finds the positive square root of a positive number. Press the square root button first, then the number, and then the equals button.

Square
A short cut to squaring a number, without having to key in the number multiplied by itself. Just key in the number then this button.

Exponent
Allows a number to be raised to a power. Enter the number, then the exponent button, then the power.

Negative
Use this to make a number negative. It is usually used when the first number in a sum is negative.

sin, cos, tan
These are mainly used in trigonometry, to find the sine, cosine, or tangent values of angles in right-angled triangles.

Brackets
These work in the same way as surrounding a part of a sum with brackets – they make sure the order of operations is correct.

△ **Scientific calculator**
A scientific calculator has many functions – a standard calculator usually only has the number pad, standard arithmetic keys, and one or two other, simpler functions, such as percentages. The keys shown here allow for more advanced maths.

Personal finance

KNOWING HOW MONEY WORKS IS IMPORTANT FOR
MANAGING YOUR PERSONAL FINANCES.

Personal finance includes paying tax on income,
gaining interest on savings, or paying interest on loans.

SEE ALSO

‹ 30–31 Positive and
negative numbers

Business finance **68–69 ›**

Formulas **169–171 ›**

Tax

Tax is a fee charged by a government on a product,
income, or activity. Governments collect the money
they need to provide services, such as schools and
defence, by taxing individuals and companies.
Individuals are taxed on what they earn – income
tax – and also on some things they buy.

GOVERNMENT
Part of the cost of
government spending
is collected in the form
of income tax

◁ **Income tax**
Each person is taxed on
what they earn – "take
home" is the amount
of money they have
left after paying their
income tax and other
deductions.

TAXPAYER
Everybody pays tax – through
their wages or through the
money that they spend

WAGE
This is the amount of money
that is earned by someone
in employment

FINANCIAL TERMS	
Financial words often seem complicated, but they are easy to understand. Knowing what the important ones mean will enable you to manage your finances by helping you to understand what you have to pay and the money you will receive.	
Bank account	This is the record of whatever a person borrows from or saves with the bank. Each account holder has a numeric password called a personal identification number (PIN), which should never be revealed to anyone.
Credit	Credit is money that is borrowed – for example, on a 4-year pay-back agreement or as an overdraft from the bank. It always costs to borrow money. The money paid to borrow from a bank is called interest.
Income	This is the money that comes to an individual or family. This can be provided by the wages that are paid for employment. Sometimes it comes from the government in the form of an allowance or direct payment.
Interest	This is the cost of borrowing money or the income received when saving with a bank. It costs more to borrow money from a bank than the interest a person would receive from the bank by saving the same amount.
Mortgage	A mortgage is an agreement to borrow money to buy a home. A bank lends the money for the purchase and this is paid back, usually over a long period of time, together with interest on the loan and other charges.
Savings	There are many forms of savings. Money can be saved in a bank to earn interest. Saving through a pension plan involves making regular payments to ensure an income after retirement.
Break-even	Break-even is the point where the cost, or what a company has spent, is equal to revenue, which is what the company has earned – at break-even the company makes neither a profit nor a loss.
Loss	Companies make a loss if they spend more than they earn – if it costs them more to produce their product than they earn by selling it.
Profit	Profit is the part of a company's income that is left once their costs have been paid – it is the money "made" by a company.

INTEREST

Banks pay interest on the money that savers invest with them (capital), and charge interest on money that is borrowed from them. Interest is given as a percentage, and there are two types, simple and compound.

Simple interest

This is interest paid only on the sum of money that is first saved with the bank. If £10,000 is put in a bank account with an interest rate of 3%, the amount will increase by the same figure each year.

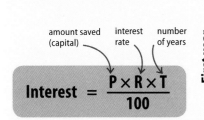

$$Interest = \frac{P \times R \times T}{100}$$

△ **Simple interest formula**
To find the simple interest made in a given year, substitute real values into this formula.

First year

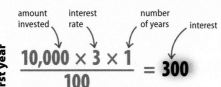

amount invested, interest rate, number of years, interest

$$\frac{10{,}000 \times 3 \times 1}{100} = 300$$

Substitute the values in the formula to work out the value of the interest for the year.

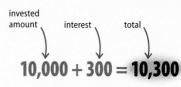

invested amount, interest, total

$$10{,}000 + 300 = 10{,}300$$

▶ **After one year**, this is the total amount of money in the saver's bank account.

Second year

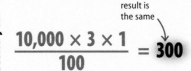

result is the same

$$\frac{10{,}000 \times 3 \times 1}{100} = 300$$

Substitute the values in the formula to work out the value of the interest for the year.

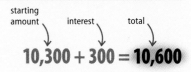

starting amount, interest, total

$$10{,}300 + 300 = 10{,}600$$

▶ **After two years** the interest is the same as the first year, as it is only paid on the initial investment.

Compound interest

This is where interest is paid on the money invested and any interest that is earned on that money. If £10,000 is paid into a bank account with an interest rate of 3%, then the amount will increase as follows.

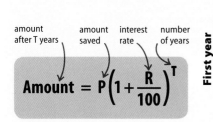

$$Amount = P\left(1 + \frac{R}{100}\right)^T$$

△ **Compound interest formula**
To find the compound interest made in a given year, substitute values into this formula.

First year

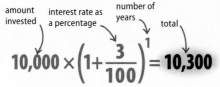

amount invested, interest rate as a percentage, number of years, total

$$10{,}000 \times \left(1 + \frac{3}{100}\right)^1 = 10{,}300$$

Substitute the values in the formula to work out the total for the first year.

total after first year, original amount, interest

$$10{,}300 - 10{,}000 = 300$$

▶ **After one year** the total interest earned is the same as that earned with simple interest (see above).

Second year

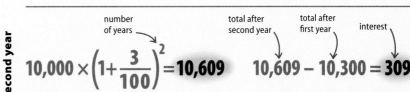

number of years

$$10{,}000 \times \left(1 + \frac{3}{100}\right)^2 = 10{,}609$$

Substitute the values in the formula to work out the total for the second year.

total after second year, total after first year, interest

$$10{,}609 - 10{,}300 = 309$$

▶ **After two years** there is a greater increase because interest is also earned on previous interest.

68 NUMBERS

Business Finance

BUSINESSES AIM TO MAKE MONEY, AND MATHS PLAYS
AN IMPORTANT PART IN ACHIEVING THIS AIM.

The aim of a business is to turn an idea
or a product into a profit, so that the
business earns more money than it
spends.

SEE ALSO	
❬ 66–67 Personal finance	
Pie charts	202–203 ❭
Line graphs	204–205 ❭

What a business does

Businesses take raw materials,
process them, and sell the end
product. To make a profit, the
business must sell its end product
at a price higher than the total
cost of the materials and the
manufacturing or production.
This example shows the basic
stages of this process using a
cake-making business.

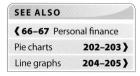

▷ **Making cakes**
This diagram shows
how a cake-making
business processes
inputs to produce
an output.

1

INPUTS

Inputs are raw materials that are
used in making a product. For
cake making, the inputs would
include the ingredients such as
flour, eggs, butter, and sugar.

◁ **Small business**
A business can consist of
just one person or a whole
team of employees.

△ **Costs**
Costs are incurred at the input stage,
as the raw materials have to be paid for.
The same costs occur every time a new
batch of cakes is made.

Revenue and profit

There is an important difference between
revenue and profit. Revenue is the money a
business makes when it sells its product. Profit
is the difference between revenue and cost –
it is the money that the business has "made".

profit is money a
business "makes"

revenue is earned by
selling final product

$$\textbf{Profit} = \textbf{Revenue} - \textbf{Costs}$$

costs are incurred in production,
for example, wages and rent

some costs are fixed, and the
money is spent however much
of the product is sold –
so costs do not start at 0

▷ **Cost graph**
This graph shows where a business
begins to make a profit: where its
revenue is greater than its costs.

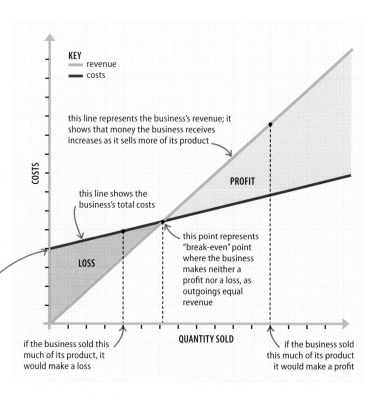

KEY
— revenue
— costs

this line represents the business's revenue; it
shows that money the business receives
increases as it sells more of its product

PROFIT

this line shows the
business's total costs

this point represents
"break-even" point
where the business
makes neither a
profit nor a loss, as
outgoings equal
revenue

LOSS

COSTS

QUANTITY SOLD

if the business sold this
much of its product, it
would make a loss

if the business sold
this much of its product
it would make a profit

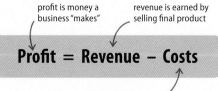

2 **PROCESSING**

Processing occurs when a business takes raw materials and turns them into something else that it can sell at a higher value.

△ **Costs**
Processing costs include rent, wages paid to staff, and the costs of utilities and equipment used for processing. These costs are often ongoing, long-term expenses.

3 **OUTPUT**

Output is what a business produces at the end of processing, in a form that is sold to customers, for example, the finished cake.

△ **Revenue**
Revenue is the money that is received by the business when it sells its output. It is used to pay off the costs. Once these are paid, the money that is left is profit.

Where the money goes

A business's revenue is not pure profit, as it must pay its costs. This pie chart shows an example of where a business's revenue might be spent, and the amount left as profit.

▷ **Costs and profit**
This pie chart shows some costs that a business might meet. Businesses that make different products have different outgoings, which reflect the make-up of their products and the efficiency of the business. When all the costs have been paid, the money left is profit.

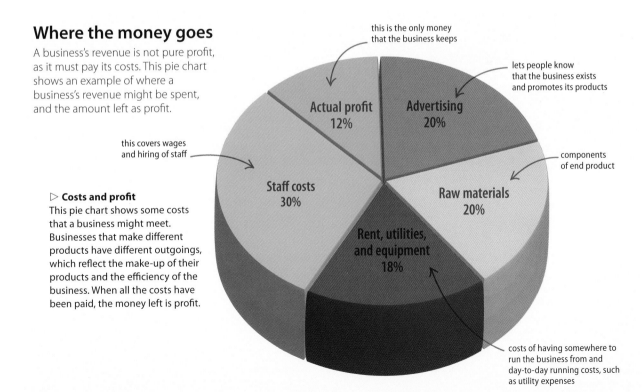

this is the only money that the business keeps

lets people know that the business exists and promotes its products

this covers wages and hiring of staff

components of end product

costs of having somewhere to run the business from and day-to-day running costs, such as utility expenses

Actual profit
12%

Advertising
20%

Staff costs
30%

Raw materials
20%

Rent, utilities, and equipment
18%

Geometry

◮ What is geometry?

GEOMETRY IS THE BRANCH OF MATHEMATICS CONCERNED
WITH LINES, ANGLES, SHAPES, AND SPACE.

Geometry has been important for thousands of years, its practical
uses include working out land areas, architecture, navigation, and
astronomy. It is also an area of mathematical study in its own right.

Lines, angles, shapes, and space

Geometry includes topics such as lines, angles, shapes (in both
two and three dimensions), areas, and volumes, but also
subjects like movements in space, such as rotations and
reflections, and coordinates.

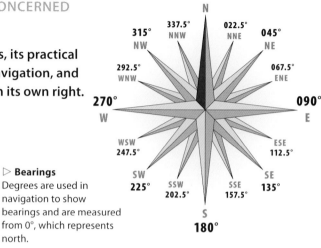

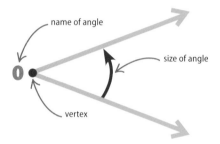

△ **Angles**
An angle is formed when two lines meet at a
point. The size of an angle is the amount of turn
between the two lines, measured in degrees.

▷ **Bearings**
Degrees are used in
navigation to show
bearings and are measured
from 0°, which represents
north.

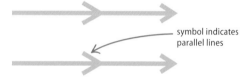

symbol indicates
parallel lines

△ **Parallel lines**
Lines that are parallel are the same
distance apart along their entire length,
and never meet, even if they are extended.

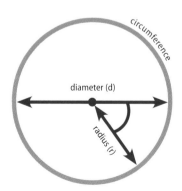

△ **Circle**
A circle is a continuous line that is always the same
distance from a central point. The length of the
line is the circumference. The diameter runs from
one side to the other through the centre. The
radius runs from the centre to the circumference.

REAL WORLD

Geometry in nature

Although many people think of geometry as a purely mathematical
subject, geometric shapes and patterns are widespread in the
natural world. Perhaps the best-known examples are the hexagonal
shapes of honeycomb cells in a beehive and of snowflakes, but
there are many other examples of natural geometry. For instance,
water droplets, bubbles, and planets are all roughly spherical.
Crystals naturally form various polyhedral
shapes – common table salt has cubic
crystals, and quartz often forms
crystals in the shape of a six-sided
prism with pyramid shaped ends.

◁ **Honeycomb cells**
Cells of honeycomb are naturally
hexagons, which can fit together
(tessellate) without leaving any
space between them.

LOOKING CLOSER

Graphs and geometry

Graphs link geometry with other areas of mathematics. Plotting lines and shapes in graphs with coordinates makes it possible to convert them into algebraic expressions, which can then be manipulated mathematically. The reverse is also true: algebraic expressions can be shown on a graph, enabling them to be manipulated using the rules of geometry. Graphical representations of objects enables positions to be given to them, which makes it possible to apply vectors and calculate the results of movements, such as rotations and translations.

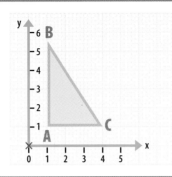

◁ **Graph**
The graph here shows a right-angled triangle, ABC, plotted on a graph. The vertices (corners) have the coordinates A = (1, 1), B = (1, 5.5), and C = (4, 1).

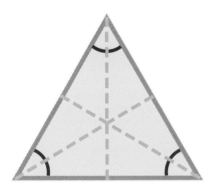

△ **Triangle**
A triangle is a three-sided, two-dimensional polygon. All triangles have three internal angles that add up to 180°.

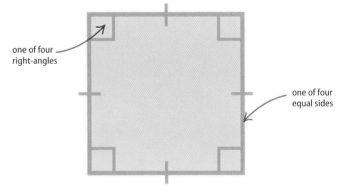

one of four right-angles

one of four equal sides

△ **Square**
A square is a four-sided polygon, or quadrilateral, in which all four sides are the same length and all four internal angles are right angles (90°).

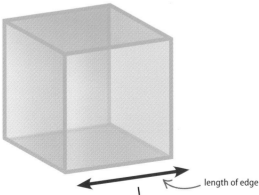

length of edge

△ **Cube**
A cube is a three-dimensional polygon in which all its edges are the same length. Like other cuboids, a cube has 6 faces, 12 edges, and 8 vertices (corners).

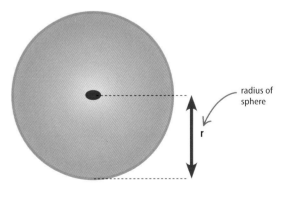

radius of sphere

r

△ **Sphere**
A sphere is a perfectly round three-dimensional shape in which every point on its surface is the same distance from the centre; this distance is the radius.

Tools in geometry

MATHEMATICAL INSTRUMENTS ARE NEEDED FOR
MEASURING AND DRAWING IN GEOMETRY.

SEE ALSO	
Angles	**76–77** ❯
Constructions	**102–105** ❯
Circles	**130–131** ❯

Tools used in geometry

Tools are vital to measure and construct geometric shapes accurately. The essential tools are a ruler, a compass, and a protractor. A ruler is used for measuring, and to draw straight lines. A compass is used to draw a whole circle or a part of a circle – called an arc. A protractor is used to measure and draw angles.

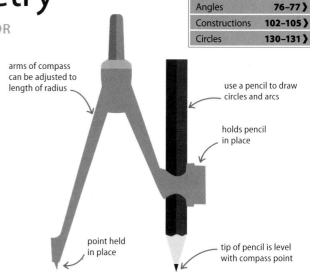

arms of compass can be adjusted to length of radius

use a pencil to draw circles and arcs

holds pencil in place

point held in place

tip of pencil is level with compass point

Using a compass

A tool for drawing circles and arcs, a compass is made up of two arms attached at one end. To use a compass, hold the arm ending in a point still, while pivoting the other arm, holding a pencil, around it. The point becomes the centre of the circle.

▽ **Drawing a circle when given the radius**
Set the distance between the arms of a compass to the given radius, then draw the circle.

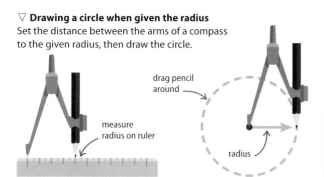

drag pencil around

measure radius on ruler

radius

Use a ruler to set the arms of the compass to the given radius.

With the compass set to the radius, hold the point down and drag the pencil around.

▽ **Drawing a circle when given its centre and one point on the circumference**
Put the point of the compass where the centre is marked and extend the other arm so that the tip of the pencil touches the point on the circumference. Then, draw the circle.

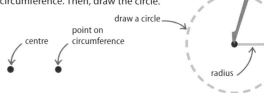

draw a circle

point on circumference

centre

radius

Set the compass to the distance between the two points.

Hold the point of the compass down and draw a circle through the point.

▽ **Drawing arcs**
Sometimes only a part of a circle – an arc – is required. Arcs are often used as guides to construct other shapes.

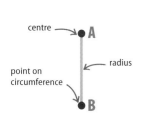

centre ● **A**

radius

point on circumference

● **B**

Draw a line and mark the ends with a point – one will be the centre of the arc, the other a point on its circumference.

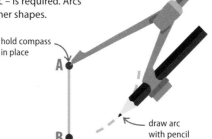

hold compass in place

A ●

B ●

draw arc with pencil

Set the compass to the length of the line – the radius of the arc – and hold it on one of the points to draw the first arc.

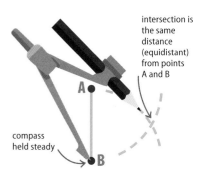

intersection is the same distance (equidistant) from points A and B

A ●

compass held steady

● **B**

Draw a second arc by holding the point of the compass on the other point. The intersection is equidistant from A and B.

Using a ruler

A ruler can be used to measure straight lines and the distances between any two points. A ruler is also necessary for setting the arms of a compass to a given distance.

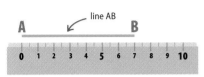

◁ **Measuring lines**
Use a ruler to measure straight lines or the distance between any two given points.

▷ **Drawing lines**
A ruler is also used as a straight edge when drawing lines between two points.

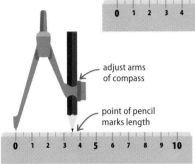

◁ **Setting a compass**
Use a ruler to measure and set the width of a compass to a given radius.

adjust arms of compass

point of pencil marks length

Other tools

Other tools may prove useful when creating drawings and diagrams in geometry.

△ **Set square**
A set square looks like a right-angled triangle and is used for drawing parallel lines. There are two types of set square, one has interior angles 90°, 40° and 45°, the other 90°, 60°, and 30.

△ **Calculator**
A calculator provides a number of key options for geometry calculations. For example, functions such as Sine can be used to work out the unknown angles of a triangle.

Using a protractor

A protractor is used to measure and draw angles. It is usually made of transparent plastic, which makes it easier to place the centre of the protractor over the point of the angle. When measuring an angle, always use the scale starting with zero.

outer scale is used to measure this obtuse angle

inner scale is used to measure this acute angle

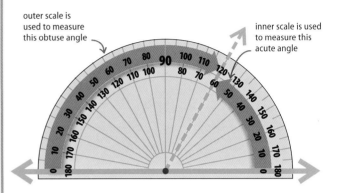

▽ **Measuring angles**
Use a protractor to measure any angle formed by two lines that meet at a point.

Extend the lines if necessary to make reading easier.

Place the protractor over the angle and read the angle measurement, making sure to read up from zero.

The other scale measures the external angle.

▽ **Drawing angles**
When given the size of an angle, use a protractor to measure and draw the angle accurately.

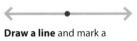

Draw a line and mark a point on it.

Place the protractor on the line with its centre over the point. Read the degrees up from zero to mark the point.

75°

Draw a line through the two points, and mark the angle.

 # Angles

AN ANGLE IS FORMED WHEN TWO LINES MEET AT A COMMON POINT.

SEE ALSO

❮ 74–75 Tools in geometry	
Straight lines	78–79 ❯
Bearings	100–101 ❯

Angles show the amount two lines "turn" as they extend in different directions away from a common point. This turn is measured in degrees, represented by the symbol °.

Measuring angles

The size of an angle depends on the amount of turn. A whole turn, making one rotation around a circle, is 360°. All other angles are less than 360°.

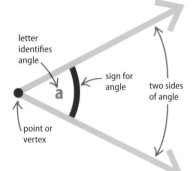

letter identifies angle
sign for angle
two sides of angle
point or vertex

△ **Parts of an angle**
The space between these two lines is the angle. An angle can be named with a letter, its value in degrees, or the symbol ∠.

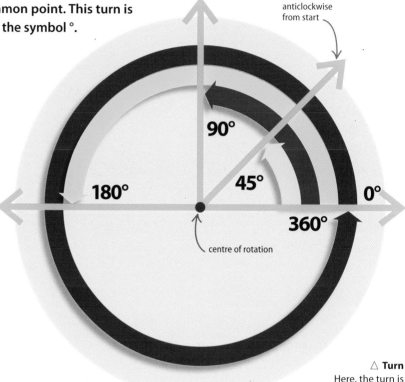

line rotated 45° anticlockwise from start

90°
180°
45°
0°
360°
centre of rotation

△ **Turn**
Here, the turn is anticlockwise, but a turn can also be clockwise.

complete turn

360°

△ **Whole turn**
An angle that is a whole turn is 360°. Such a rotation brings both sides of the angle back to the starting point.

half turn of circle
straight line

180°

△ **Half turn**
An angle that is a half turn is 180°. Its two sides form a straight line. The angle is also known as a straight angle.

quarter turn

90°

△ **Quarter turn**
An angle that is a quarter turn is 90°. Its two sides are perpendicular (L-shaped). It is also known as a right angle.

45°

△ **Eighth turn**
An angle that is one eighth of a whole turn is 45°. It is half of a right angle, and eight of these angles are a whole turn.

Types of angle

There are four important types of angle, which are shown below. They are named according to their size.

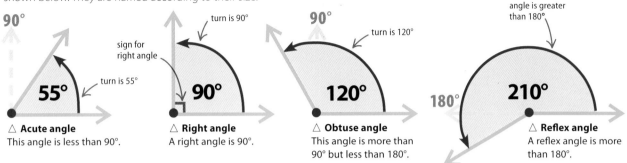

△ **Acute angle**
This angle is less than 90°.

△ **Right angle**
A right angle is 90°.

△ **Obtuse angle**
This angle is more than 90° but less than 180°.

△ **Reflex angle**
A reflex angle is more than 180°.

Naming angles

Angles can have individual names and names that reflect a shared relationship.

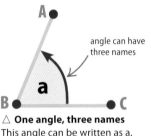

△ **One angle, three names**
This angle can be written as a, or as ∠ ABC, or as ∠ CBA.

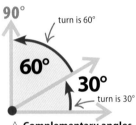

△ **Complementary angles**
Any two angles that add up to 90° are complementary.

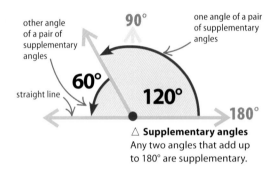

△ **Supplementary angles**
Any two angles that add up to 180° are supplementary.

Angles on a straight line

The angles on a straight line make up a half turn and so they add up to 180°. In this example, four adjacent angles add up to the 180° of a straight line.

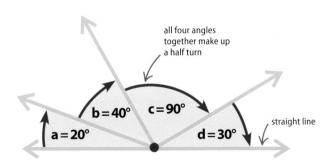

$$a+b+c+d=180°$$
$$20°+40°+90°+30°=180°$$

Angles at a point

The angles surrounding a point, or vertex, make up a whole turn and so they add up to 360°. In this example, five adjacent angles at the same point add up to the 360° of a complete circle.

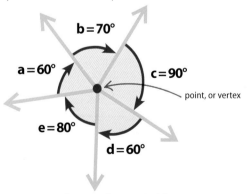

$$a+b+c+d+e=360°$$
$$60°+70°+90°+60°+80°=360°$$

Straight lines

A STRAIGHT LINE IS USUALLY JUST CALLED A LINE. IT IS THE SHORTEST
DISTANCE BETWEEN TWO POINTS ON A SURFACE OR IN SPACE.

SEE ALSO

‹ 72–73 Tools in geometry
‹ 74–75 Angles
Constructions **102–105 ›**

Points, lines, and planes

The most fundamental objects in geometry are points, lines,
and planes. A point represents a specific position and has no
width, height, or length. A line is one dimensional – it has
infinite length extending in two opposite directions. A plane
is a two-dimensional flat surface extending in all directions.

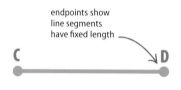

△ **Points**
A point is used to represent a precise
location. It is represented by a dot
and named with a capital letter.

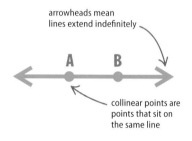

△**Lines**
A line is represented by a straight line and
arrowheads signify that it extends indefinitely
in both directions. It can be named by any two
points that it passes through – this line is AB.

△ **Line segments**
A line segment has fixed length, so it will
have endpoints rather than arrowheads. A
line segment is named by its endpoints –
this is line segment CD.

△ **Planes**
A plane is usually represented by a two-
dimensional figure and labelled with a capital
letter. Edges can be drawn, although a plane
actually extends indefinitely in all directions.

Sets of lines

Two lines on the same surface, or plane,
can either intersect – meaning they
share a point – or they can be parallel.
If two lines are the same distance apart
along their lengths and never intersect,
they are parallel.

△ **Non-parallel lines**
Non parallel lines are not the same distance
apart all the way along; if they are extended
they will eventually meet in a point.

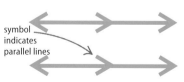

△ **Parallel lines**
Parallel lines are two or more lines that never
meet, even if extended. Identical arrows are
used to indicate lines that are parallel.

△ **Transversal**
Any line that intersects two or more
other lines, each at a different point,
is called a transversal.

LOOKING CLOSER

Parallelograms

A parallelogram is a four-sided shape
with two pairs of opposite sides, both
parallel and of equal length.

△ **Parallel sides**
The sides AB and DC are parallel, as are
sides BC and AD. The sides AB and BC,
and AD and CD are not parallel – shown
by the different arrows on these lines.

Angles and parallel lines

Angles can be grouped and named according to their relationships with straight lines. When parallel lines are crossed by a transversal, it creates pairs of equal angles – each pair has a different name.

▽ Labelling angles
Lines AB and CD are parallel. The angles created by the intersecting transversal line are labelled with lower-case letters.

arrows indicate lines AB and CD are parallel

all angles with one arc in this diagram are the same

transversal line crosses parallel lines

all angles with two arcs in this diagram are the same

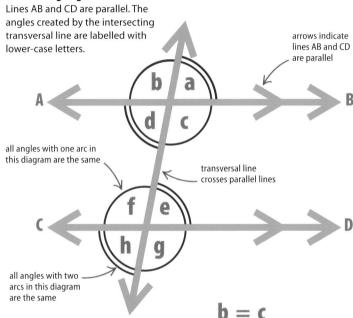

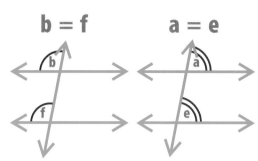

b = f   **a = e**

△ Corresponding angles
Angles in the same position in relation to the transversal line and one of a pair of parallel lines, are called corresponding angles. These angles are equal.

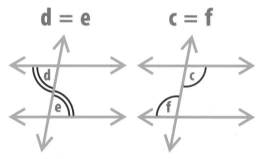

d = e   **c = f**

△ Alternate angles
Alternate angles are formed on either side of a transversal between parallel lines. These angles are equal.

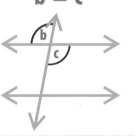

b = c

▷ Vertically-opposite angles
When two lines cross, equal angles are formed on opposite sides of the point. These angles are known as vertically-opposite angles.

Drawing a parallel line

Drawing a line that is parallel to an existing line requires a pencil, a ruler, and a protractor.

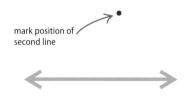

mark position of second line

measure this angle between the original line and the line that crosses through the point

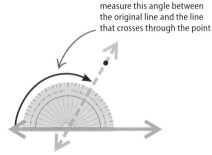

these angles are equal

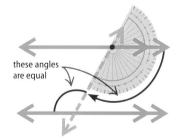

Draw a straight line with a ruler. Mark a point – this will be the distance of the new, parallel line from the original line.

Draw a line through the mark, intersecting the original line. This is the transversal. Measure the angle it makes with the original line.

Measure the same angle from the transversal. Draw the new line through the mark with a ruler; this line is parallel to the original line.

Symmetry

THERE ARE TWO TYPES OF SYMMETRY –
REFLECTIVE AND ROTATIONAL.

SEE ALSO
❮ 78–79 Straight lines
Rotations **92–93 ❯**
Reflections **94–95 ❯**

A shape has symmetry when a line can be drawn that splits the shape exactly into two, or when it can fit into its outline in more than one way.

Reflective symmetry

A flat (two-dimensional) shape has reflective symmetry when each half of the shape on either side of a bisecting line (mirror line) is the mirror image of the other half. This mirror line is called a line of symmetry.

▽ **Lines of symmetry**
These are the lines of symmetry for some flat or two-dimensional shapes. Circles have an unlimited number of lines of symmetry.

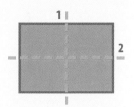

Lines of symmetry of a rectangle

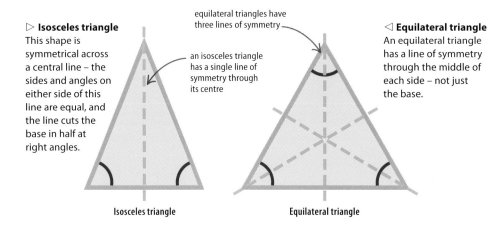

▷ **Isosceles triangle**
This shape is symmetrical across a central line – the sides and angles on either side of this line are equal, and the line cuts the base in half at right angles.

equilateral triangles have three lines of symmetry

an isosceles triangle has a single line of symmetry through its centre

◁ **Equilateral triangle**
An equilateral triangle has a line of symmetry through the middle of each side – not just the base.

Isosceles triangle

Equilateral triangle

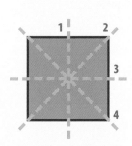

Lines of symmetry of a square

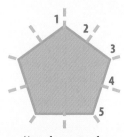

Lines of symmetry of a regular pentagon

Planes of symmetry

Solid (three-dimensional) shapes can be divided using "walls" known as planes. Solid shapes have reflective symmetry when the two sides of the shape split by a plane are mirror images.

▽ **Cuboid**
Formed by three pairs of rectangles, a cuboid can be divided into two symmetrical shapes in three ways.

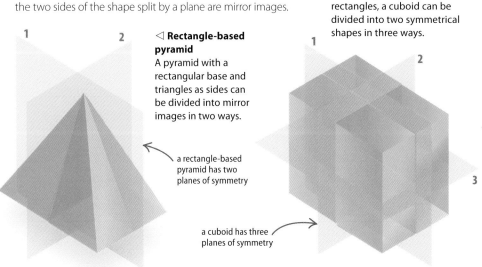

◁ **Rectangle-based pyramid**
A pyramid with a rectangular base and triangles as sides can be divided into mirror images in two ways.

a rectangle-based pyramid has two planes of symmetry

a cuboid has three planes of symmetry

Every line through the middle of a circle is a line of symmetry

Rotational symmetry

A two-dimensional shape has rotational symmetry when it can be rotated about a point, called the centre of rotation, and still exactly fit its original outline. The number of ways it fits its outline when rotated is called its "order" of rotational symmetry.

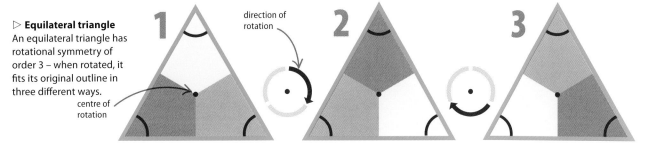

▷ **Equilateral triangle**
An equilateral triangle has rotational symmetry of order 3 – when rotated, it fits its original outline in three different ways.

direction of rotation

centre of rotation

▽ Square
A square has rotational symmetry of order 4 – when rotated around its centre of rotation, it fits its original outline in four different ways.

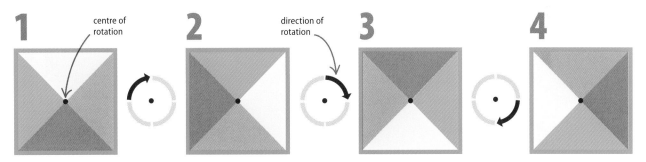

centre of rotation

direction of rotation

Axes of symmetry

Instead of a single point as the centre of rotation, a three-dimensional shape is rotated around a line known as its axis of symmetry. It has rotational symmetry if, when rotated, it fits into its original outline.

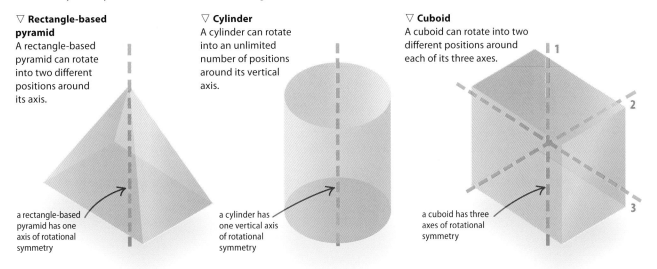

▽ **Rectangle-based pyramid**
A rectangle-based pyramid can rotate into two different positions around its axis.

a rectangle-based pyramid has one axis of rotational symmetry

▽ **Cylinder**
A cylinder can rotate into an unlimited number of positions around its vertical axis.

a cylinder has one vertical axis of rotational symmetry

▽ **Cuboid**
A cuboid can rotate into two different positions around each of its three axes.

a cuboid has three axes of rotational symmetry

Coordinates

COORDINATES GIVE THE POSITION OF A PLACE OR POINT ON A MAP OR GRAPH.

SEE ALSO

Vectors	86–89 ❯
Linear graphs	174–177 ❯

Introducing coordinates

Coordinates come in pairs of numbers or letters, or both. They are always written in brackets separated by a comma. The order in which coordinates are read and written is important. In this example, (E, 1), means four units, or squares on this map, to the right (along the horizontal row) and one square down, or up in some cases, (the vertical column).

▽ **City map**
A grid provides a framework for locating places on a map. Every square is identified by two coordinates. A place is found when the horizontal coordinate meets the vertical coordinate. On this city map, the horizontal coordinates are letters and the vertical coordinates are numbers. On other maps only numbers may be used.

numbers are used as vertical coordinates on this map

letters are used as horizontal coordinates on this map

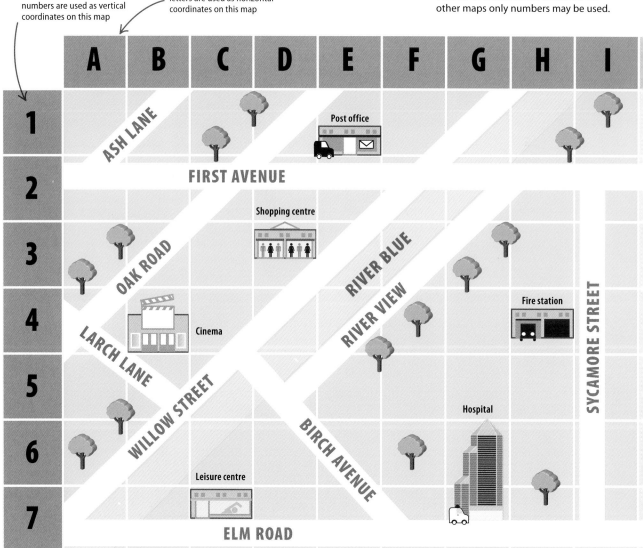

Map reading

The horizontal coordinate is always given first and the vertical coordinate second. On the map below, a letter and a number are paired together to form a coordinate.

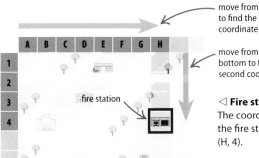

move from left to right to find the first coordinate

move from top to bottom to find the second coordinate

◁ **Fire station**
The coordinates of the fire station are (H, 4).

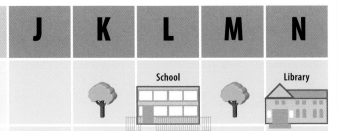

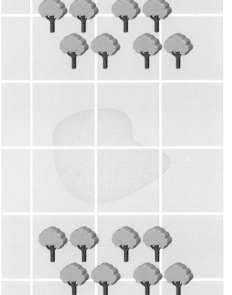

FIRST AVENUE

School

Library

MAPLE SQUARE

Town hall

ELM ROAD

Using coordinates

Each place of interest on this map can be found using the given coordinates. Remember when reading this map to first read across (horizontal) and then down (vertical).

◁ **Cinema**
Find the cinema using coordinates (B, 4). Start from square A and move 1 square to the right, then move 4 squares down.

◁ **Post office**
The coordinates of the post office are (E, 1). Find the horizontal coordinate E then move down 1 square.

◁ **Town hall**
Find the town hall using coordinates (J, 5). From square A, move 9 squares to the right, then move 5 squares down.

◁ **Leisure centre**
Using the coordinates (C, 7), find the location of the leisure centre. First, find C. Next, find 7 on the vertical column.

◁ **Library**
The coordinates of the library are (N, 1). Find N first then move down 1 square to locate the library.

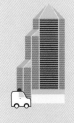

◁ **Hospital**
The hospital can be found using the coordinates (G, 7). To find the horizontal coordinate of G, move 6 squares to the right. Then go down 6 squares to find the vertical coordinate 7.

◁ **Fire station**
Find the fire station using coordinates (H, 4). Move 7 squares to the right to find H, then move 4 squares down.

◁ **School**
The coordinates of the school are (L, 1). First find L, then move down 1 square to find the school.

◁ **Shopping centre**
Using the coordinates (D, 3), find the location of the shopping centre. Find D. Next, find 3 on the vertical column.

Graph coordinates

Coordinates are used to identify the positions of points on graphs, in relation to two axes – the y axis is a vertical line, and the x axis a horizontal line. The coordinates of a point are written as its position on the x axis, followed by its position on the y axis, (x, y).

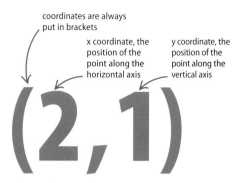

coordinates are always put in brackets

x coordinate, the position of the point along the horizontal axis

y coordinate, the position of the point along the vertical axis

▷ **Four quadrants**
Coordinates are measured on axes, which cross at a point called the "origin". These axes create four quadrants. There are positive values on the axes above and to the right of the origin, and negative values below and to its left.

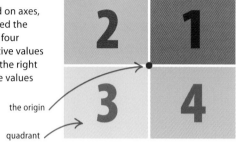

the origin

quadrant

△ **Coordinates of a point**
Coordinates give the position of a point on each axis. The first number gives its position on the x axis, the second its position on the y axis.

Plotting coordinates

Coordinates are plotted on a set of axes. To plot a given point, first read along to its value on the x axis, then read up or down to its value on the y axis. The point is plotted where the two values cross each other.

$$A = (2, 2) \quad B = (-1, -3)$$
$$C = (1, -2) \quad D = (-2, 1)$$

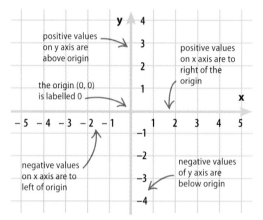

positive values on y axis are above origin

positive values on x axis are to right of the origin

the origin (0, 0) is labelled 0

negative values on x axis are to left of origin

negative values of y axis are below origin

These are four sets of coordinates. Each gives its x value first, followed by its y value. Plot the points on a set of axes.

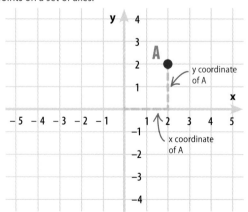

y coordinate of A

x coordinate of A

▶ **To plot each point,** look at its x coordinate (the first number), and read along the x axis from 0 to this number. Then read up or down to its y coordinate (the second number).

Using squared paper, draw a horizontal line to form the x axis, and a vertical line to be the y axis. Number the axes, with the origin separating the positive and negative values.

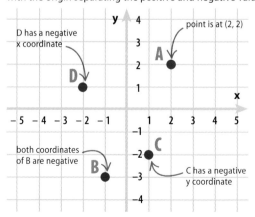

D has a negative x coordinate

point is at (2, 2)

both coordinates of B are negative

C has a negative y coordinate

▶ **Plot each point in the same way.** With negative coordinates, the process is the same, but read to the left instead of right for an x coordinate, and down instead of up for a y coordinate.

Equation of a line

Lines that pass through a set of coordinates on a pair of axes can be expressed as equations. For example, on the line of the equation $y = x + 1$, any point that lies on the line has a y coordinate that is 1 unit greater than its x coordinate.

y coordinate

x coordinate

$$y = x + 1$$

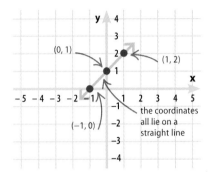

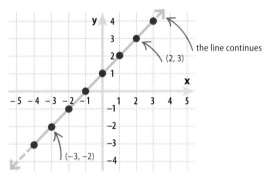

the line continues

The equation of a line can be found using only a few coordinates. This line passes through the coordinates (–1, 0), (0, 1), and (1, 2), so it is already clear what pattern the points follow.

The graph of the equation is of all the points where the y coordinate is 1 greater than the x coordinate ($y = x + 1$). This means that the line can be used to find other coordinates that satisfy the equation.

World map

Coordinates are used to mark the position of places on the Earth's surface, using lines of latitude and longitude. These work in the same way as the x and y axes on a graph. The "origin" is the point where the Greenwich Meridian (0 for longitude) crosses the Equator (0 for latitude).

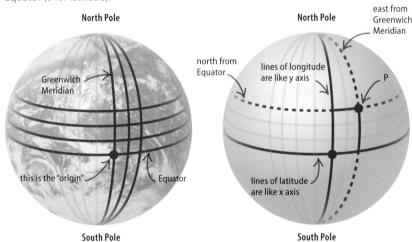

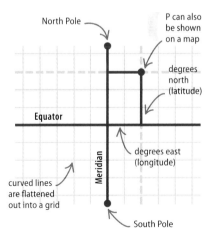

Lines of longitude run from the North Pole to the South Pole. Lines of latitude are at right-angles to lines of longitude. The origin is where the Equator (x axis) crosses the Greenwich Meridian (y axis).

The coordinates of a point such as P are found by finding how many degrees East it is from the Meridian and how many degrees North it is from the Equator.

This is how the surface of the Earth is shown on a map. The lines of latitude and longitude work the same way as axes –the vertical lines show latitude and horizontal lines show longitude.

Vectors

A VECTOR IS A LINE THAT HAS SIZE (MAGNITUDE) AND DIRECTION.

A vector is a way to show a distance in a particular direction. It is often drawn as a line with an arrow on it. The length of the line shows the size of the vector and the arrow gives its direction.

SEE ALSO

❬ **82–85** Coordinates

Translations **90–91** ❭

Pythagoras' theorem **120–121** ❭

intended direction of travel

direction of current

horizontal direction

starting point

actual direction of travel

end point swimmer is aiming for

vertical direction

vector is determined by other two lines

end point

What is a vector?

A vector is a distance in a particular direction. Often, a vector is a diagonal distance, and in these cases it forms the diagonal side (hypotenuse) of a right-angled triangle (see pp.120–121). The other sides of the triangle determine the vector's length and direction. In the example on the left, a swimmer's path is a vector. The other two sides of the triangle are the distance across to the opposite shore from the starting point, and the distance down from the end point that the swimmer was aiming for to the actual end point where the swimmer reaches the shore.

◁ **Vector of a swimmer**

A man sets out to swim to the opposite shore of a river that is 30m wide. A current pushes him as he swims, and he ends up 20m downriver from where he intended. His path is a vector with dimensions 30 across and 20 down.

Expressing vectors

In diagrams, a vector is drawn as a line with an arrow on it, showing its size and direction. There are three different ways of writing vectors using letters and numbers.

$\mathbf{v} =$
A "v" is a general label for a vector, used even when its size is known. It is often used as a label in a diagram.

$\overrightarrow{\mathbf{ab}} =$
Another way of representing a vector is by giving its start and end points, with an arrow above them to show direction.

$\begin{pmatrix} 6 \\ 4 \end{pmatrix} =$
The size and direction of the vector can be shown by giving the horizontal units over the vertical units.

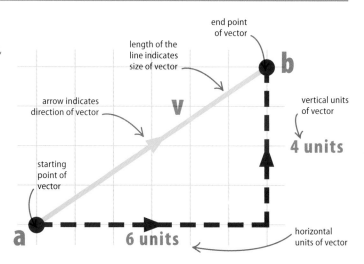

end point of vector

length of the line indicates size of vector

arrow indicates direction of vector

v

b

vertical units of vector

4 units

starting point of vector

a

6 units

horizontal units of vector

Direction of vectors

The direction of a vector is determined by whether its units are positive or negative. Positive horizontal units mean movement to the right, negative horizontal units mean left; positive vertical units mean movement up, and negative vertical units mean down.

▷ **Movement up and left**
This movement has a vector with negative horizontal units and positive vertical units.

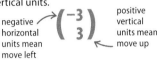

negative horizontal units mean move left

positive vertical units mean move up

$\begin{pmatrix} -3 \\ 3 \end{pmatrix}$

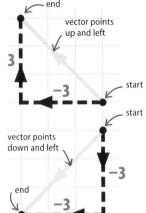

end
vector points up and left
3
−3
start

▷ **Movement up and right**
This movement has a vector with both sets of units positive.

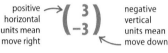

positive horizontal units mean move right

positive vertical units mean move up

$\begin{pmatrix} 3 \\ 3 \end{pmatrix}$

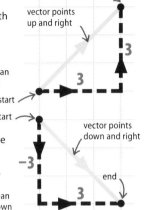

end
vector points up and right
3
3
start

▷ **Movement down and left**
This movement has a vector with both sets of units negative.

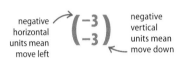

negative horizontal units mean move left

negative vertical units mean move down

$\begin{pmatrix} -3 \\ -3 \end{pmatrix}$

vector points down and left
start
−3
end
−3

▷ **Movement down and right**
This movement has a vector with positive horizontal units and negative vertical units.

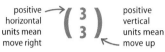

positive horizontal units mean move right

negative vertical units mean move down

$\begin{pmatrix} 3 \\ -3 \end{pmatrix}$

start
vector points down and right
−3
end
3

Equal vectors

Vectors can be identified as equal even if they are in different positions on the same grid, as long as their horizontal and vertical units are equal.

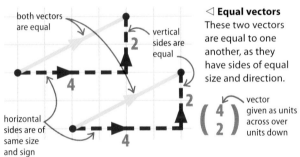

both vectors are equal
4
2
horizontal sides are of same size and sign
4
2

◁ **Equal vectors**
These two vectors are equal to one another, as they have sides of equal size and direction.

vertical sides are equal

$\begin{pmatrix} 4 \\ 2 \end{pmatrix}$

vector given as units across over units down

▷ **Equal vectors**
These two vectors are equal to one another because their horizontal and vertical sides are both the same size and have the same direction.

$\begin{pmatrix} -1 \\ -5 \end{pmatrix}$

numerical expression of both vectors

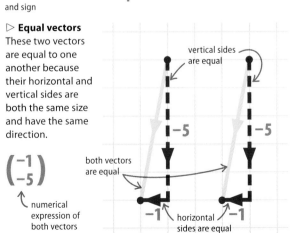

vertical sides are equal
both vectors are equal
−5
−5
−1
−1
horizontal sides are equal

Magnitude of vectors

With diagonal vectors, the vector is the longest side (c) of a right-angled triangle. Use Pythagoras' theorem to find the length of a vector from its vertical (a) and horizontal (b) units.

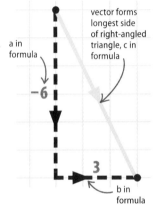

a in formula
vector forms longest side of right-angled triangle, c in formula
−6
3
b in formula

formula for Pythagoras' theorem

$$a^2 + b^2 = c^2$$

Put the vertical and horizontal units of the vector into the formula.

$$-6^2 + 3^2 = c^2$$

$-6^2 = -6 \times -6 = 36$

$3^2 = 3 \times 3 = 9$

$$36 + 9 = c^2$$

Find the squares by multiplying each value by itself.

c^2 is the square of vector

$$45 = c^2$$

Add the two squares. This total equals c^2 (the square of the vector).

square root of 45

c is length of vector

$$c = \sqrt{45}$$

Find the square root of the total value (45) by using a calculator.

length of vector

$$c = 6.7$$

The answer is the magnitude (length) of the vector.

Adding and subtracting vectors

Vectors can be added and subtracted in two ways. The first is by using written numbers to add the horizontal and vertical values. The second is by drawing the vectors end to end, then seeing what new vector is created.

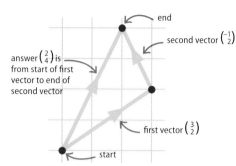

answer $\binom{2}{4}$ is from start of first vector to end of second vector

second vector $\binom{-1}{2}$

first vector $\binom{3}{2}$

start

end

▷ **Addition**
Vectors can be added in two different ways. Both methods give the same answer.

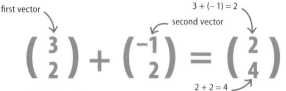

first vector

second vector

$3 + (-1) = 2$

$$\binom{3}{2} + \binom{-1}{2} = \binom{2}{4}$$

$2 + 2 = 4$

△ **Adding the parts**
To add vectors numerically, add the two top numbers (the horizontal values) and then the two bottom numbers (the vertical values).

△ **Addition by drawing vectors**
Draw one vector, then draw the second starting from the end point of the first. The answer is the new vector that has been created, from the start of the first vector to the end of the second.

▷ **Subtraction**
Vectors can be subtracted in two different ways. Both methods give the same answer.

first vector, from which second vector is subtracted

second vector, which is subtracted from first vector

$3 - (-1) = 4$

$$\binom{3}{2} - \binom{-1}{2} = \binom{4}{0}$$

$2 - 2 = 0$

△ **Subtracting the parts**
To subtract one vector from another, subtract its vertical value from the vertical value of the first vector, then do the same for the horizontal values.

first vector $\binom{3}{2}$

to subtract vectors, second vector $\binom{-1}{2}$ is reversed, giving $\binom{1}{-2}$

start

end

answer $\binom{4}{0}$ is from start of first vector to end of second vector

△ **Subtraction by drawing vectors**
Draw the first vector, then draw the second vector reversed, starting from the end point of the first vector. The answer to the subtraction is the vector from the start point to the end point.

Multiplying vectors

Vectors can be multiplied by numbers, but not by other vectors. The direction of a vector stays the same if it is multiplied by a positive number, but is reversed if it is multiplied by a negative number. Vectors can be multiplied by drawing or by using their numerical values.

▽ **Vector a**
Vector a has −4 horizontal units and +2 vertical units. It can be shown as a written vector or a drawn vector, as shown below.

$$\mathbf{a} = \binom{-4}{2}$$

horizontal value

vertical value

end

drawn vector

vector a

2

−4

start

▽ **Vector a multiplied by 2**
To multiply vector a by 2 numerically, multiply each of its horizontal and vertical parts by 2. To multiply it by 2 by drawing, simply extend the original vector by the same length again.

vector a

$2 \times -4 = -8$

$$2\mathbf{a} = 2 \times \binom{-4}{2} = \binom{-8}{4}$$

$2 \times 2 = 4$

end

$2 \times 2 = 4$ vertical units

4

$2 \times -4 = -8$ horizontal units

vector 2a is twice as long as vector a

start

−8

▽ **Vector a multiplied by −½**
To multiply vector a by −½ numerically, multiply each of its parts by −½. To multiply it by −½ by drawing, draw a vector half the length and in the opposite direction to a.

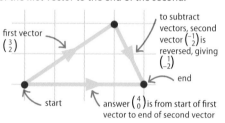

vector a

$-\frac{1}{2} \times -4 = +2$

$$-\frac{1}{2}\mathbf{a} = -\frac{1}{2} \times \binom{-4}{2} = \binom{+2}{-1}$$

$-\frac{1}{2} \times 2 = -1$

$2 \times -\frac{1}{2} = -1$ vertical units

vector −½a is in the opposite direction

−1

2

$-4 \times -\frac{1}{2} = +2$ horizontal units

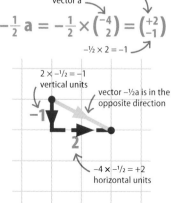

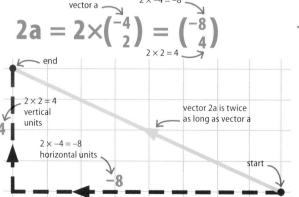

Working with vectors in geometry

Vectors can be used to prove results in geometry. In this example, vectors are used to prove that the line joining the midpoints of any two sides of a triangle is parallel to the third side of the triangle, as well as being half its length.

First, choose 2 sides of triangle ABC, in this example AB and AC. Mark these sides as the vectors a and b. To get from B to C, go along BA and then AC, rather than BC. BA is the vector –a as it is the opposite of AB, and AC is just b. This means vector BC is –a + b.

$$\vec{BC} = -\,a + b$$

vector BC

this is negative because BA is opposite of AB

vector BC can also be expressed like this

$$-\,a + b$$

vector AB is labelled a

$$a$$

vector AC is labelled b

$$b$$

Second, find the midpoints of the two sides that have been chosen (AB and AC). Mark the midpoint of AB as P, and the midpoint of AC as Q. This creates three new vectors: AP, AQ, and PQ. AP is half the length of vector a, and AQ is half the length of vector b.

$$\vec{AP} = \tfrac{1}{2}\,\vec{AB} = \tfrac{1}{2}\,a$$

$$\vec{AQ} = \tfrac{1}{2}\,\vec{AC} = \tfrac{1}{2}\,b$$

P is the midpoint of AB

$$\tfrac{1}{2}\,a$$

$$\tfrac{1}{2}\,b$$

Q is the midpoint of AC

Third, use the vectors ½a and ½b to find the length of vector PQ. To get from P to Q go along PA then AQ. PA is the vector –½a, as it is the opposite of AP, and AQ is already known to be ½b. This means vector PQ is –½a + ½b.

this is half BA

this is half AC

$$\vec{PQ} = -\tfrac{1}{2}\,a + \tfrac{1}{2}\,b$$

$$\vec{PQ} = \tfrac{1}{2}\,\vec{BC}$$

BC is –a + b, so PQ is half BC

this is negative because BA is opposite of AB

$$-\tfrac{1}{2}\,a$$

$$-\,a + b$$

$$-\tfrac{1}{2}\,a + \tfrac{1}{2}\,b$$

PQ is half BC

$$\tfrac{1}{2}\,b$$

Fourth, make the proof. The vectors PQ and BC are in the same direction and are therefore parallel to each other, so the line PQ (which joins the midpoints of the sides AB and AC) must be parallel to the line BC. Also, vector PQ is half the length of vector BC, so the line PQ must be half the length of the line BC.

BC and PQ are parallel

vector BC

$$-\,a + b$$

vector PQ

$$-\tfrac{1}{2}\,a + \tfrac{1}{2}\,b$$

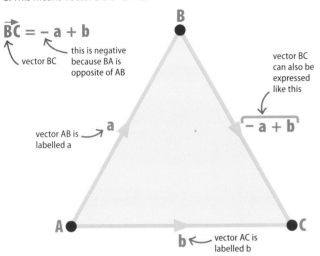
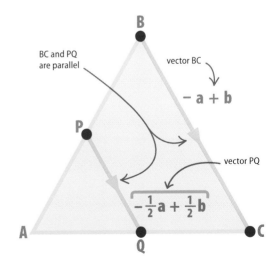

Translations

A TRANSLATION CHANGES THE POSITION OF A SHAPE.

A translation is a type of transformation. It moves an object to a new position.
The translated object is called an image, and it is exactly the same size and
shape as the original object. Translations are written as vectors.

SEE ALSO

‹ 82–85 Coordinates
‹ 86–89 Vectors
Rotations   92–93 ›
Reflections   94–95 ›
Enlargements   96–97 ›

How translations work

A translation moves an object to a new position, without making any other changes, for example to size or shape.
Here, the triangle named ABC is translated so that its image is the triangle A₁B₁C₁. This translation is named T₁. The
triangle A₁B₁C₁ is then translated again and its image is the triangle A₂B₂C₂. This second translation is named T₂.

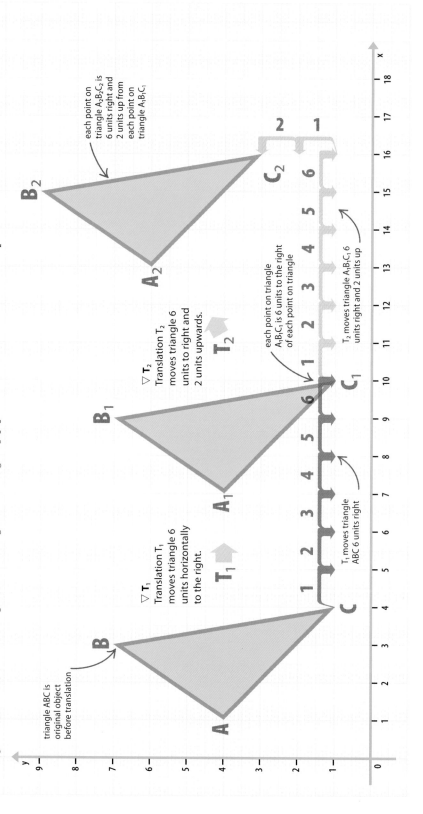

triangle ABC is
original object
before translation

▽ T₁
Translation T₁
moves triangle 6
units horizontally
to the right.

T₁ moves triangle
ABC 6 units right

▽ T₂
Translation T₂
moves triangle 6
units to right and
2 units upwards.

each point on triangle
A₁B₁C₁ is 6 units to the right
of each point on triangle

T₂ moves triangle A₁B₁C₁ 6
units right and 2 units up

each point on
triangle A₂B₂C₂ is
6 units right and
2 units up from
each point on
triangle A₁B₁C₁

Writing translations

Translations are written as vectors. The top number shows the horizontal distance an object moves, while the bottom number shows the vertical distance moved. The two numbers are contained within a set of brackets. Each translation can be numbered – for example T_1, T_2, T_3 – to make it clear which one is being referred to if more than one translation is shown.

translation number

$$T_1 = \begin{pmatrix} 6 \\ 0 \end{pmatrix}$$

distance moved horizontally

distance moved vertically

△ **Translation T_1**
To move triangle ABC to position $A_1B_1C_1$ each point moves 6 units horizontally, but does not move vertically. The vector is written as above.

translation number

$$T_2 = \begin{pmatrix} 6 \\ 2 \end{pmatrix}$$

distance moved horizontally

distance moved vertically

△ **Translation T_2**
To move triangle $A_1B_1C_1$ to position $A_2B_2C_2$ each point moves 6 units horizontally, then moves 2 units vertically. The vector is written as above.

Direction of translations

The numbers used to show a translation's vector are positive or negative, depending on which direction the object moved. If it moves to the right or up, it is positive; to the left or down, it is negative.

▽ **Negative translation**
The rectangle ABCD, moves down and left, so the values in its vector are negative.

this is the original shape

moves 1 unit down

this is the translation T_1

moves 3 units to left

▽ **Translation T_1**
The translation T_1, moves rectangle ABCD to the new position $A_1B_1C_1D_1$. It is written as the vector shown – both its parts are negative.

$$T_1 = \begin{pmatrix} -3 \\ -1 \end{pmatrix}$$

distance moved horizontally (to the left)

distance moved vertically (downwards)

Tessellations in action

A tessellation is a pattern created using shapes to cover a surface without leaving any gaps. Two shapes can be tessellated with themselves using only translation (and no rotation) – the square and the regular hexagon. To tessellate a hexagon using translation requires 6 different translations; to tessellate a square requires 8.

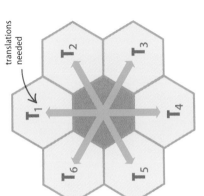

translations needed

△ **Hexagons**
Each of the hexagons around the outside is a translated image of the central hexagon. The tessellation continues in the same way.

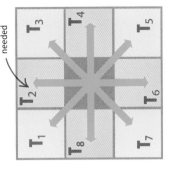

translations needed

△ **Squares**
Each of the squares around the edge is a translated image of the central square. The tessellation continues in the same way.

Rotations

A ROTATION IS A TYPE OF TRANSFORMATION THAT TAKES AN OBJECT AND MOVES IT ABOUT A GIVEN POINT.

SEE ALSO

❰ 76–77 Angles
❰ 82–85 Coordinates
❰ 90–91 Translations
Reflections   94–95 ❱
Enlargements   96–97 ❱
Constructions   102–105 ❱

The point around which a rotation occurs is called the centre of rotation, the distance a shape turns is called the angle of rotation.

Properties of a rotation

Rotations occur around a fixed point called the centre of rotation, and are measured by angles. Any point on an original object and the corresponding point on its rotated image will be exactly the same distance from the centre of rotation. The centre of rotation can sit inside, outside, or on the outline of an object. A rotation can be drawn, or the centre and angle of an existing rotation found, using a compass, ruler, and protractor.

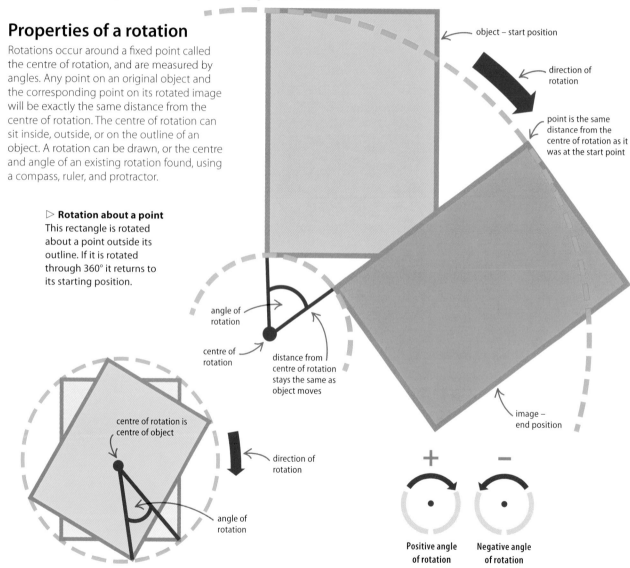

▷ **Rotation about a point**
This rectangle is rotated about a point outside its outline. If it is rotated through 360° it returns to its starting position.

object – start position

direction of rotation

point is the same distance from the centre of rotation as it was at the start point

angle of rotation

centre of rotation

distance from centre of rotation stays the same as object moves

image – end position

centre of rotation is centre of object

direction of rotation

angle of rotation

+

−

Positive angle of rotation

Negative angle of rotation

△ **Rotation about a point inside an object**
An object can be rotated about a point that is inside it rather than outside – this rectangle has been rotated around its centre point. It will fit into its outline again if it rotates through 180°.

△ **Angle of rotation**
The angle of rotation is either positive or negative. If it is positive, the object rotates in a clockwise direction; if it is negative, it rotates anticlockwise.

Construction of a rotation

To construct a rotation, three elements of information are needed: the object to be rotated, the location of the centre of rotation, and the size of the angle of rotation.

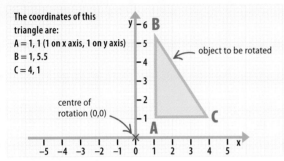

The coordinates of this triangle are:
A = 1, 1 (1 on x axis, 1 on y axis)
B = 1, 5.5
C = 4, 1

centre of rotation (0,0)

object to be rotated

Given the position of the triangle ABC (see above) and the centre of rotation, rotate the triangle −90°, which means 90° anticlockwise. The image of triangle ABC will be on the left-hand side of the y axis.

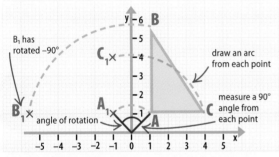

B_1 has rotated −90°

draw an arc from each point

measure a 90° angle from each point

angle of rotation

Place a compass point on the centre of rotation and draw arcs anticlockwise from points A, B, and C (anticlockwise because the rotation is negative). Then, placing the centre of a protractor over the centre of rotation, measure an angle of 90° from each point. Mark the point where the angle meets the arc.

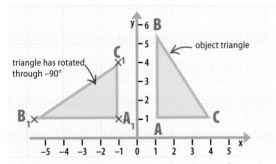

object triangle

triangle has rotated through −90°

Label the new points A_1, B_1, and C_1. Join them to form the image. Each point on the new triangle $A_1B_1C_1$ has rotated 90° anticlockwise from each point on the original triangle ABC.

Finding the angle and centre of a rotation

Given an object and its rotated image, the centre and angle of rotation can be found.

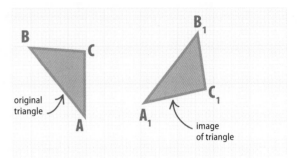

original triangle

image of triangle

The triangle $A_1B_1C_1$ is the image of triangle ABC after a rotation. The centre and angle of rotation can be found by drawing the perpendicular bisectors (lines that cut exactly through the middle – see pp.102–103) of the lines between two sets of points, here A and A_1 and B and B_1.

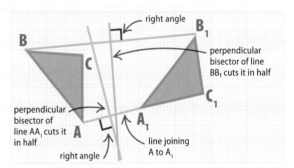

right angle

perpendicular bisector of line BB_1 cuts it in half

perpendicular bisector of line AA_1 cuts it in half

line joining A to A_1

right angle

Using a compass and a ruler, construct the perpendicular bisector of the line joining A and A_1 and the perpendicular bisector of the line that joins B and B_1. These bisectors will cross each other.

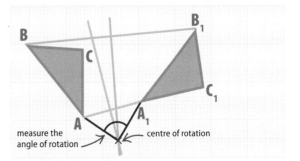

measure the angle of rotation

centre of rotation

The centre of rotation is the point where the two perpendicular bisectors cross. To find the angle of rotation, join A and A_1 to the centre of rotation and measure the angle between these lines.

Reflections

A REFLECTION SHOWS AN OBJECT TRANSFORMED INTO ITS
MIRROR IMAGE ACROSS AN AXIS OF REFLECTION.

SEE ALSO

❰ **80–81** Symmetry
❰ **82–85** Coordinates
❰ **90–91** Translations
❰ **92–93** Rotations
Enlargements **96–97** ❱

Properties of a reflection

Any point on an object (for example, A) and the
corresponding point on its reflected image (for
example, A_1) are on opposite sides of, and equal
distances from, the axis of reflection. The reflected
image is effectively a mirror image whose base sits
along the axis of reflection.

△ **Reflected mountain**
The mountain on which the points
A, B, C, D, and E are marked has a
reflected image, which includes
the points A_1, B_1, C_1, D_1, and E_1.

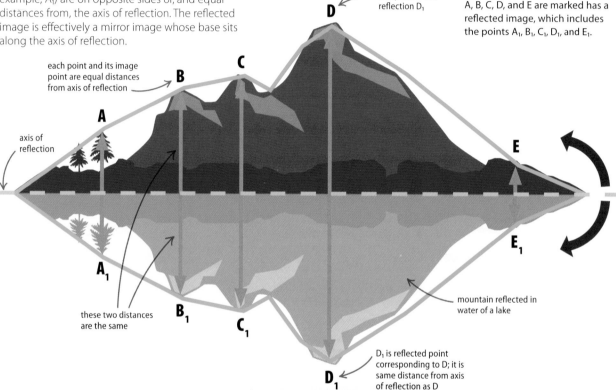

each point and its image
point are equal distances
from axis of reflection

axis of
reflection

this point D has
reflection D_1

D

C

B

A

E

E_1

these two distances
are the same

A_1

B_1

C_1

D_1

mountain reflected in
water of a lake

D_1 is reflected point
corresponding to D; it is
same distance from axis
of reflection as D

Kaleidoscopes

A kaleidoscope creates patterns using mirrors and coloured beads. The patterns are
the result of beads being reflected and then reflected again.

two mirrors

this is one
reflection of the
original beads

the final
reflection, which
completes image

A simple kaleidoscope contains two
mirrors at right angles (90°) to each
other and some coloured beads.

The beads are reflected in the
two mirrors, producing two reflected
images on either side.

Each of the two reflections is
reflected again, producing another
image of the beads.

Constructing a reflection

To construct the reflection of an object it is necessary to know the position of the axis of reflection and of the object. Each point on the reflection will be the same distance from the axis of reflection as its corresponding point on the original. Here, the reflection of triangle ABC is drawn for the axis of reflection y = x (which means that each point on the axis has the same x and y coordinates).

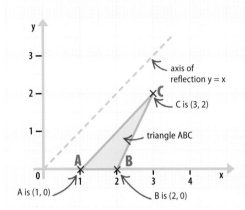

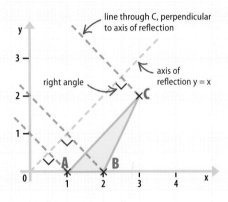

First, draw the axis of reflection. As y = x, this axis line crosses through the points (0, 0), (1, 1), (2, 2), (3, 3), and so on. Then draw in the object that is to be reflected – the triangle ABC, which has the coordinates (1, 0), (2, 0), and (3, 2). In each set of coordinates, the first number is the x value, and the second number is the y value.

Second, draw lines from each point of the triangle ABC that are at right-angles (90°) to the axis of reflection. These lines should cross the axis of reflection and continue onwards, as the new coordinates for the reflected image will be measured along them.

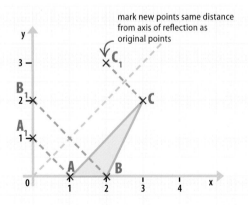

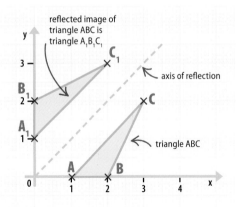

Third, measure the distance from each of the original points to the axis of reflection, then measure the same distance on the other side of the axis to find the positions of the new points. Mark each of the new points with the letter it reflects, followed by a small 1, for example A_1.

Finally, join the points A_1, B_1, and C_1 to complete the image. Each of the points of the triangle has a mirror image across the axis of reflection. Each point on the original triangle is an equal distance from the axis of reflection as its reflected point.

Enlargements

AN ENLARGEMENT IS A TRANSFORMATION THAT TAKES AN OBJECT AND
PRODUCES AN IMAGE OF THE SAME SHAPE BUT OF DIFFERENT SIZE.

SEE ALSO
❮ **48–51** Ratio and proportion
❮ **90–91** Translations
❮ **92–93** Rotations
❮ **94–95** Reflections

Enlargements are constructed through a fixed point known as the centre
of enlargement. The image can be larger or smaller. The change in size is
determined by a number called the scale factor.

Properties of an enlargement

When an object is transformed into a larger image, the
relationship between the corresponding sides of that
object and the image is the same as the scale factor.
For example, if the scale factor is 5, the sides of the
image are 5 times bigger than those of the original.

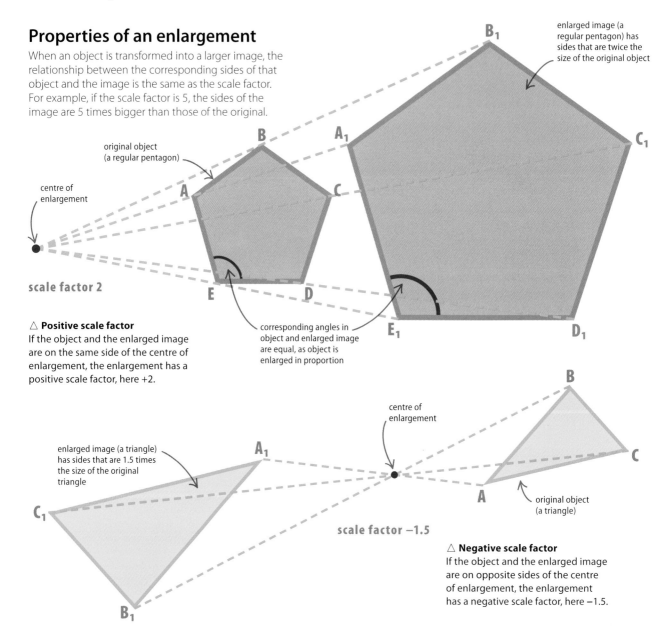

enlarged image (a regular pentagon) has sides that are twice the size of the original object

original object (a regular pentagon)

centre of enlargement

scale factor 2

corresponding angles in object and enlarged image are equal, as object is enlarged in proportion

△ **Positive scale factor**
If the object and the enlarged image
are on the same side of the centre of
enlargement, the enlargement has a
positive scale factor, here +2.

enlarged image (a triangle) has sides that are 1.5 times the size of the original triangle

centre of enlargement

scale factor −1.5

original object (a triangle)

△ **Negative scale factor**
If the object and the enlarged image
are on opposite sides of the centre
of enlargement, the enlargement
has a negative scale factor, here −1.5.

Constructing an enlargement

An enlargement is constructed by plotting the coordinates of the object on squared (or graph) paper. Here, the quadrilateral ABCD is measured through the centre of enlargement (0, 0) with a given scale factor of 2.5.

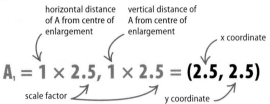

horizontal distance of A from centre of enlargement

vertical distance of A from centre of enlargement

x coordinate

$$A_1 = 1 \times 2.5, 1 \times 2.5 = (2.5, 2.5)$$

scale factor

y coordinate

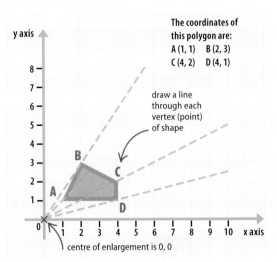

The coordinates of this polygon are:
A (1, 1)   B (2, 3)
C (4, 2)   D (4, 1)

draw a line through each vertex (point) of shape

centre of enlargement is 0, 0

The same principle is then applied to the other points, to work out their x and y coordinates.

$$B_1 = 2 \times 2.5, 3 \times 2.5 = (5, 7.5)$$

$$C_1 = 4 \times 2.5, 2 \times 2.5 = (10, 5)$$

$$D_1 = 4 \times 2.5, 1 \times 2.5 = (10, 2.5)$$

▷ **Draw the polygon** ABCD using the given coordinates. Mark the centre of enlargement and draw lines from this point through each of the vertices of the shape (points where sides meet).

▷ **Then, calculate the positions** of A_1, B_1, C_1, and D_1 by multiplying the horizontal and vertical distances of each point from the centre of enlargement (0, 0) by the scale factor 2.5.

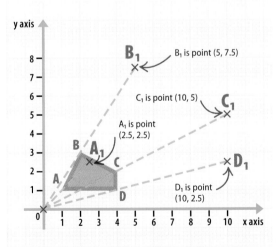

B_1 is point (5, 7.5)

C_1 is point (10, 5)

A_1 is point (2.5, 2.5)

D_1 is point (10, 2.5)

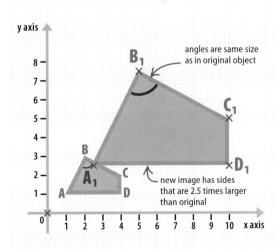

angles are same size as in original object

new image has sides that are 2.5 times larger than original

▷ **Read along the x axis and the y axis** to plot the vertices (points) of the enlarged image. For example, B_1 is point (5, 7.5) and C_1 is point (10, 5). Mark and label all the points A_1, B_1, C_1, and D_1.

▷ **Join the new coordinates** to complete the enlargement. The enlarged image is a quadrilateral with sides that are 2.5 times larger than the original object, but with angles of exactly the same size.

Scale drawings

A SCALE DRAWING SHOWS AN OBJECT ACCURATELY AT A PRACTICAL SIZE BY REDUCING OR ENLARGING IT.

Scale drawings can be scaled down, such as a map, or scaled up, such as a diagram of a microchip.

SEE ALSO

⟨ **48–51** Ratio and proportion
⟨ **96–97** Enlargements
Circles   **130–131** ⟩

scale shows how lengths on bridge are reduced in drawing

Scale: ←

1cm : 10m

convert scale to ratio of 1cm : 1,000cm using centimetres as common unit

Choosing a scale

To make an accurate plan of a large object, such as a bridge, the object's measurements need to be scaled down. To do this, every measurement of the bridge is reduced by the same ratio. The first step in creating a scale drawing is to choose a scale – for example, 1cm for each 10m. The scale is then shown as a ratio, using the smallest common unit.

length (in cm) on scale drawing

length (in cm) of real length

1cm : 1,000cm

symbol for ratio

◁ **Scale as a ratio**
A scale of 1cm to 10m can be shown as a ratio by using centimetres as a common unit. There are 100cm in a metre, so $10 \times 100cm = 1{,}000cm$.

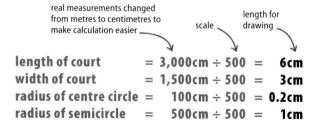

60m

How to make a scale drawing

In this example, a basketball court needs to be drawn to scale. The court is 30m long and 15m wide. In its centre is a circle with a radius of 1m, and at either end a semicircle, each with a radius of 5m. To make a scale drawing, first make a rough sketch, noting the real measurements. Next, work out a scale. Use the scale to convert the measurements, and create the final drawing using these.

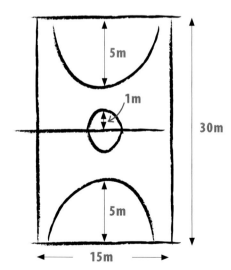

5m

1m

30m

5m

15m

Seeing as 30m (the longest length of the drawing) needs to fit into a space of less than 10cm, a convenient scale is chosen:

measurement on drawing

1cm : 5m

measurement on real court

By converting this to a ratio of 1cm : 500cm, it is now possible to work out the measurements that will be used in the drawing.

real measurements changed from metres to centimetres to make calculation easier

scale

length for drawing

length of court	=	3,000cm ÷ 500	=	**6cm**
width of court	=	1,500cm ÷ 500	=	**3cm**
radius of centre circle	=	100cm ÷ 500	=	**0.2cm**
radius of semicircle	=	500cm ÷ 500	=	**1cm**

Draw a rough sketch to act as a guide, marking on it the real measurements. Make a note of the longest length (30m). Based on this and the space available for your drawing, work out a suitable scale.

▷ **Choose a suitable scale** and convert it into a ratio by using the lowest common unit, centimetres. Next, convert the real measurements into the same units. Divide each measurement by the scale to find the measurements for the drawing.

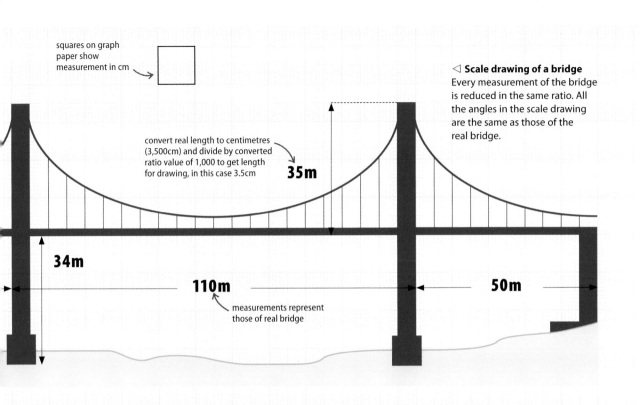

squares on graph paper show measurement in cm →

◁ **Scale drawing of a bridge**
Every measurement of the bridge is reduced in the same ratio. All the angles in the scale drawing are the same as those of the real bridge.

convert real length to centimetres (3,500cm) and divide by converted ratio value of 1,000 to get length for drawing, in this case 3.5cm

35m

34m

110m

measurements represent those of real bridge

50m

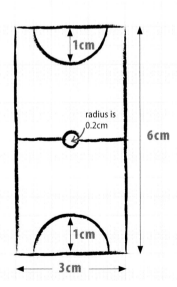

1cm

radius is 0.2cm

6cm

1cm

3cm

▷ **Make a second rough sketch,** this time marking on the scaled measurements. This provides a guide for the final drawing.

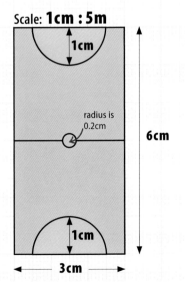

Scale: **1cm : 5m**

1cm

radius is 0.2cm

6cm

1cm

3cm

▷ **Construct a final**, accurate scale drawing of the basketball court. Use a ruler to draw the lines, and a compass to draw the circle and semicircles.

REAL WORLD

Maps

The scale of a map varies according to the area it covers. To see a whole country such as France a scale of 1cm : 150km might be used. To see a town a scale of 1cm : 500m is suitable.

⬆ Bearings

A BEARING IS A WAY OF SHOWING A DIRECTION.

Bearings show accurate directions. They can be used to plot journeys through unfamiliar territory, where it is vital to be exact.

SEE ALSO

❬ **74–75** Tools in geometry

❬ **76–77** Angles

❬ **98–99** Scale drawings

What are bearings?

Bearings are angles measured clockwise from the compass direction north. They are usually given as three-digit whole numbers of degrees, such as 270°, but they can also use decimal numbers, such as with 247.5°. Compass directions are given in terms such as "WSW", or "west-south-west".

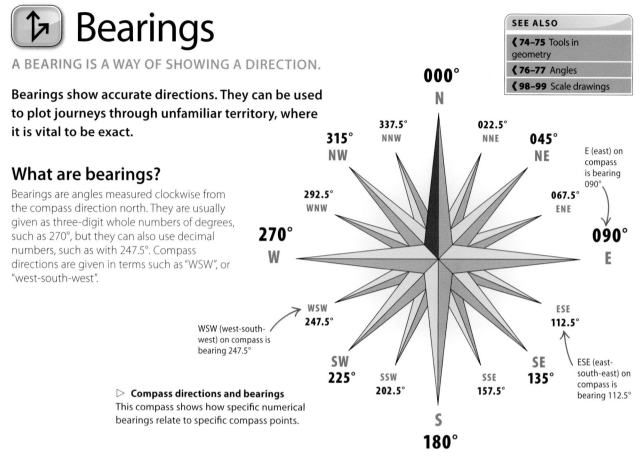

WSW (west-south-west) on compass is bearing 247.5°

E (east) on compass is bearing 090°

ESE (east-south-east) on compass is bearing 112.5°

▷ **Compass directions and bearings**
This compass shows how specific numerical bearings relate to specific compass points.

How to measure a bearing

Begin by deciding on the start point of the journey. Place a protractor at this start or centre point. Use the protractor to draw the angle of the bearing clockwise from the compass direction north.

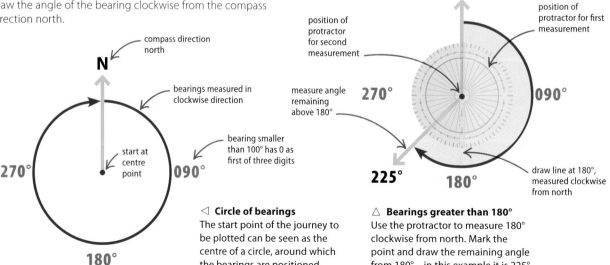

compass direction north

bearings measured in clockwise direction

start at centre point

bearing smaller than 100° has 0 as first of three digits

position of protractor for second measurement

measure angle remaining above 180°

position of protractor for first measurement

draw line at 180°, measured clockwise from north

◁ **Circle of bearings**
The start point of the journey to be plotted can be seen as the centre of a circle, around which the bearings are positioned.

△ **Bearings greater than 180°**
Use the protractor to measure 180° clockwise from north. Mark the point and draw the remaining angle from 180° – in this example it is 225°.

Plotting a journey with bearings

Bearings are used to plot journeys of several direction changes. In this example, a plane flies on the bearing 290° for 300km, then turns to the bearing 045° for 200km. Plot its last leg back to the start, using a scale of 1cm for 100km.

1cm : 100km

First, draw the bearing 290°.
Set the protractor at the centre and draw 180°. Draw a further 110°, giving a total of 290°.

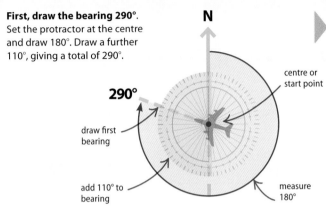

290°

draw first bearing

add 110° to bearing

measure 180°

centre or start point

Second, work out the distance travelled on the bearing 290°. Using the scale, the distance is 3cm, because 1cm equals 100km.

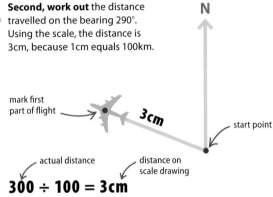

N

mark first part of flight

3cm

start point

actual distance   distance on scale drawing

300 ÷ 100 = 3cm

Set the protractor at the end point of the 3cm and draw a new north line. The next bearing is 045° from this north.

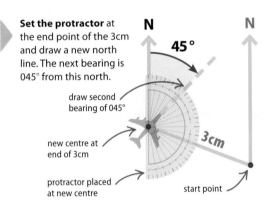

N   N

45°

draw second bearing of 045°

new centre at end of 3cm

3cm

protractor placed at new centre

start point

Work out the distance travelled on the bearing 045°. Using the scale of 1cm for every 100km, the distance is 2cm.

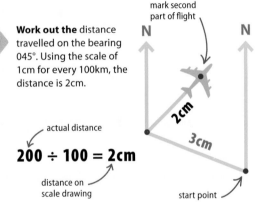

mark second part of flight

N   N

2cm

3cm

actual distance

200 ÷ 100 = 2cm

distance on scale drawing

start point

Set the protractor at the end point of the 2cm and draw a new north line. The next bearing is found to be 150° from this latest north. This direction takes the plane back to the start point.

x = 150°

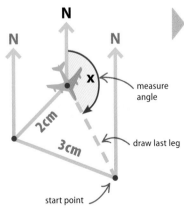

N

N   N

x

measure angle

2cm

3cm

draw last leg

start point

Finally, draw the distance travelled on the bearing 150°. Using the scale, the distance is 2.8cm, meaning the final leg of the journey is 280km.

y = 2.8cm

distance on scale drawing

2.8 × 100 = 280km

actual distance of last leg of journey

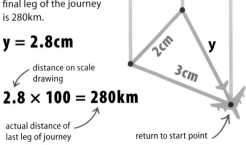

N

N   N

2cm   y

3cm

return to start point

Constructions

MAKING PERPENDICULAR LINES AND ANGLES USING
A COMPASS AND A STRAIGHT EDGE.

SEE ALSO

〈 74-75 Tools in Geometry

〈 76-77 Angles

Triangles **108–109 〉**

Congruent Triangles **112–113 〉**

An accurate geometric drawing is called a construction.
These drawings can include lines, angles, and shapes. The
tools needed are a compass and a straight edge.

Constructing perpendicular lines

Two lines are perpendicular when they intersect (or cross) at 90°,
or right angles. There are two ways to construct a perpendicular line
– the first is to draw through a point marked on a given line, and the
second is to use a point that is above or below the given line.

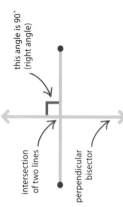

this angle is 90°
(right angle)

intersection
of two lines

perpendicular
bisector

▽ **Perpendicular bisector**
A perpendicular bisector cuts
another line exactly in half,
crossing through its midpoint
at right angles, or 90°.

Using a point on the line

A perpendicular line can be constructed by using a point marked on a line. The
point marked is where the two lines will intersect (cross) at right angles.

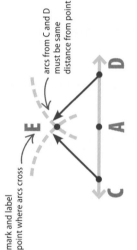

mark a point
roughly halfway
along the line

each arc will
cross the line

arcs must be
same distance
from point A

mark and label points
where arcs cross the line

Draw a line and mark a point on the line with a
letter, for example, A. Place the point of a
compass on point A, and draw two arcs of the
same distance from either side of this point.

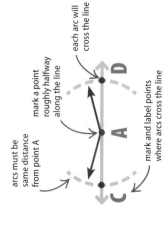

mark and label
point where arcs cross

arcs from C and D
must be same
distance from point

Place the point of a compass on point
C, and draw an arc above the line. Do the
same from point D. The arcs will intersect
(cross) at a point, label this point E.

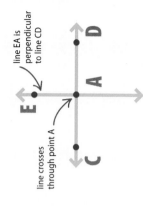

line EA is
perpendicular
to line CD

line crosses
through point A

▶ **Now, draw a line** from E through A.
This line is perpendicular (at right
angles) to the original line.

Using a point above the line

Perpendicular lines can be constructed by marking a point above the first line, through which the second perpendicular line will pass.

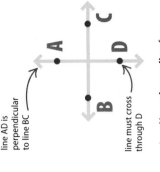

draw a line

mark point above line → A

Draw a line and mark a point above it. Label this mark with a letter, for example, A.

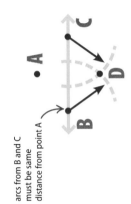

draw two arcs from A

arcs from A cross line at two points

▸ **Place a compass** on point A. Draw two arcs that intersect the line at two points. Label these points B and C.

arcs from B and C must be same distance from point A

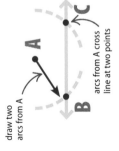

▸ **With the compass** on points B and C, draw two arcs of the same length beneath the line. The two arcs will intersect at a point, label this point D.

line AD is perpendicular to line BC

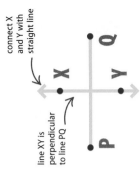

line must cross through D

▸ **Now, draw a line** from points A to D. This line is perpendicular (at right angles) to line BC.

Constructing a perpendicular bisector

A line that passes exactly through the midpoint of a line segment at right angles, or 90°, is called a perpendicular bisector. It can be constructed by marking points above and below the line segment.

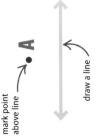

line PQ

First, draw a line, and label each end point, for example, P and Q.

draw arc from point P

set compass length to just over half of line PQ

▸ **Place a compass** on point P and draw an arc with a distance just over half the length of line PQ.

arcs from Q will cross arcs from P

keep compass at same distance

place compass on point Q

▸ **Draw another arc** from point Q with the compass kept at the same length. This arc will intersect the first arc at two points.

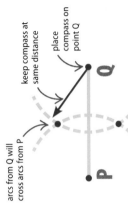

connect X and Y with straight line

line XY is perpendicular to line PQ

▸ **Label the points** where the arcs intersect X and Y. Draw a line connecting X and Y; this is the perpendicular bisector of line PQ.

Bisecting an angle

The bisector of an angle is a straight line that intersects the vertex (point) of the angle, splitting it into two equal parts. This line can be constructed by using a compass to mark points on the sides of the angle.

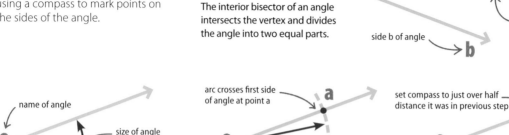

▷ **An angle bisector**
The interior bisector of an angle intersects the vertex and divides the angle into two equal parts.

side a of angle

angle bisector cuts through centre of vertex (point)

bisector is halfway between sides a and b

side b of angle

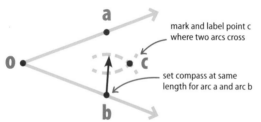

name of angle

size of angle

vertex

First, draw an angle of any size. Label the vertex of this angle with a letter, for example, o.

arc crosses first side of angle at point a

point b is where arc crosses second side of angle

Draw an arc by placing the point of a compass on the vertex. Mark and label the points at which the arc intersects the angle's sides.

set compass to just over half distance it was in previous step

Place the compass on point a and draw an arc in the space between the angle's sides.

mark and label point c where two arcs cross

set compass at same length for arc a and arc b

Keep the compass set at the same length and place it on point b, then draw another arc. The two arcs intersect at a point, label this intersection c.

line that joins vertex and point c is angle bisector

Draw a line from the vertex, o, through point c – this is the angle bisector. The angle is now split into two equal parts.

Congruent triangles

Triangles are congruent if all their sides and interior angles are equal. The points that are marked when drawing an angle bisector create two congruent triangles – one above the bisector and one below.

▷ **Constructing triangles**
By connecting the points made after drawing a bisecting line through an angle, two congruent triangles are formed.

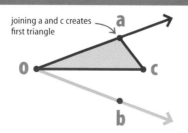

joining a and c creates first triangle

Draw a line from a to c to make the first triangle – shaded red here.

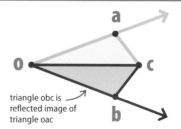

triangle obc is reflected image of triangle oac

Now, draw a line from b to c to construct the second triangle – shaded red here.

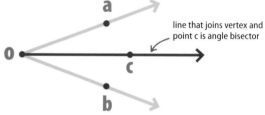

Constructing 90° and 45° angles

Bisecting an angle can be used to construct some common angles without using a protractor, for example a right angle (90°) and a 45° angle.

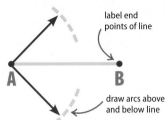

label end points of line

label points where arcs cross

draw arcs above and below line

Draw a straight line (AB). Place a compass on point A, set it to a distance just over half of the line's length, and draw an arc above and below the line.

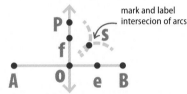

compass must be set to same size as arcs from point A

Then, draw two arcs with the compass set to the same length and placed on point B. Label the points where the arcs cross each other P and Q.

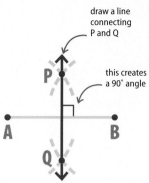

draw a line connecting P and Q

this creates a 90° angle

Draw a line from point P to point Q. This is a perpendicular bisector of the original line and it creates four 90° angles.

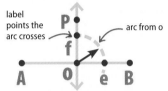

label points the arc crosses

arc from o

Draw an arc from point o that crosses two lines on either side, this creates a 90° angle. Label the two points f and e where the arc intersects the lines.

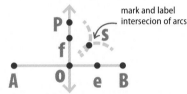

mark and label intersecion of arcs

Keep the compass at the same length as the last arc and draw arcs from points f and e. Label the intersection of these arcs with a letter (s).

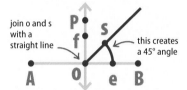

join o and s with a straight line

this creates a 45° angle

Draw a line from point o through s. This line is the angle bisector. The 90° angle is now split into two 45° angles.

Constructing 60° angles

An equilateral triangle, which has three equal sides and three 60° angles, can be constructed without a protractor.

label line with letters

first line can be any length

2.5cm

Draw a line, which will form one arm of the first angle. Here the line is 2.5cm long, but it can be any length. Mark each end of the line with a letter.

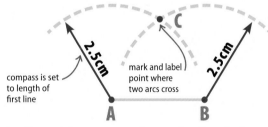

compass is set to length of first line

mark and label point where two arcs cross

2.5cm

2.5cm

Now, set the compass to the same length as the first line. Draw an arc from point A, then another from point B. Mark the point where the two arcs cross C.

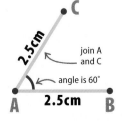

join A and C

angle is 60°

2.5cm

2.5cm

Now, draw a line to connect points A and C. Line AC is the same length as line AB. A 60° angle has been created.

Construct an equilateral triangle by drawing a third line from B to C. Each side of the triangle is equal and each internal angle of the triangle is 60°.

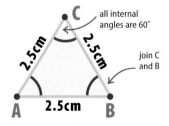

all internal angles are 60°

join C and B

2.5cm

2.5cm

2.5cm

Loci

A LOCUS (PLURAL LOCI) IS THE PATH FOLLOWED BY A POINT
THAT ADHERES TO A GIVEN RULE WHEN IT MOVES.

SEE ALSO

❰ **74–75** Tools in geometry
❰ **98–99** Scale drawings
❰ **102–105** Constructions

What is a locus?

Many familiar shapes, such as circles, and straight lines are examples of
loci because they are paths of points that conform to specific conditions.
Loci can also produce more complicated shapes. They are often used to
solve practical problems, for example, pinpointing an exact location.

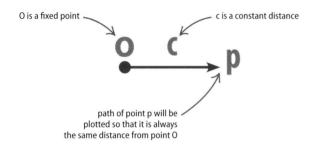

O is a fixed point

c is a constant distance

path of point p will be
plotted so that it is always
the same distance from point O

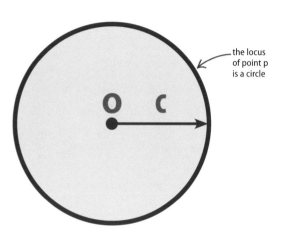

the locus
of point p
is a circle

A compass and a pencil are needed to construct this
locus. The point of the compass is held in the fixed point,
O. The arms of the compass are spread so that the
distance between its arms is the constant distance, c.

The shape drawn when turning the compass a full
rotation reveals that the locus is a circle. The centre
of the circle is O and the radius is the fixed distance
between the compass point and the pencil (c).

Working with loci

To draw a locus, it is necessary to find all the points that conform to the rule
that has been specified. This will require a compass, a pencil, and a ruler. This
example shows how to find the locus of a point that moves so that its
distance from a fixed line AB is always the same.

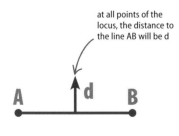

at all points of the
locus, the distance to
the line AB will be d

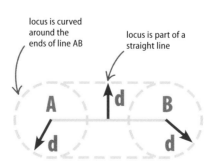

locus is curved
around the
ends of line AB

locus is part of a
straight line

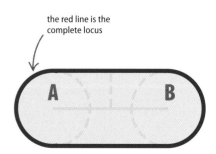

the red line is the
complete locus

Draw the line segment AB. A and B are
fixed points. Now, plot the distance of d
from the line AB.

Between points A and B, the locus is a straight
line. At the end of these lines, the locus is a
semicircle. Use a compass to draw these.

This is the completed locus.
It has the shape of a typical
athletics track.

Spiral locus

Loci can follow more complex paths. The example below follows the path of a piece of string that is wound around a cylinder, creating a spiral locus.

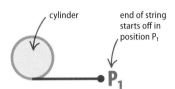

cylinder

end of string starts off in position P$_1$

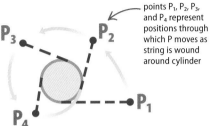

points P$_1$, P$_2$, P$_3$, and P$_4$ represent positions through which P moves as string is wound around cylinder

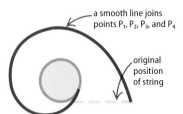

a smooth line joins points P$_1$, P$_2$, P$_3$, and P$_4$

original position of string

▶ **The string starts off** lying flat, with point P$_1$ the position of the end of the string.

▶ **As the string** is wound around the cylinder, the end of the string moves closer to the surface of the cylinder.

▶ **When the path** of point P is plotted, it forms a spiral locus.

Using loci

Loci can be used to solve difficult problems. Suppose two radio stations, A and B, share the same frequency, but are 200km apart. The range of their transmitters is 150km. The area where the ranges of the two transmitters overlap, or interference, can be found by showing the locus of each transmitter and using a scale drawing (see pp.98–99).

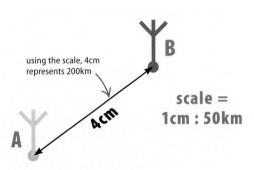

using the scale, 4cm represents 200km

scale = 1cm : 50km

4cm

To find the area of interference, first choose a scale, then draw the reach of each transmitter. An appropriate scale for this example is 1cm : 50km.

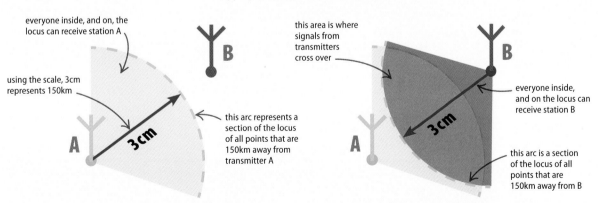

everyone inside, and on, the locus can receive station A

using the scale, 3cm represents 150km

this arc represents a section of the locus of all points that are 150km away from transmitter A

3cm

this area is where signals from transmitters cross over

everyone inside, and on the locus can receive station B

this arc is a section of the locus of all points that are 150km away from B

3cm

▶ **Construct the reception area** for radio station A. Draw the locus of a point that is always 150km from station A. The scale gives 150km = 3cm, so draw an arc with a radius of 3cm, with A as the centre.

▶ **Construct the reception area** for radio station B. This time draw an arc with the compass set to 3cm, with B as the centre. The interference occurs in the area where the two paths overlap.

Triangles

A TRIANGLE IS A SHAPE FORMED WHEN THREE STRAIGHT LINES MEET.

A triangle has three sides and three interior angles. A vertex (plural vertices) is the point where two sides of a triangle meet. A triangle has three vertices.

SEE ALSO

❰ 76–77 Angles
❰ 78–79 Straight lines
Constructing triangles   110–111 ❱
Polygons   126–129 ❱

Introducing triangles

A triangle is a three-sided polygon. The base of a triangle can be any one of its three sides, but it is usually the bottom one. The longest side of a triangle is opposite the biggest angle. The shortest side of a triangle is opposite the smallest angle. The three interior angles of a triangle add up to 180°.

shortest side

longest side

▲ABC

biggest angle

smallest angle

△ **Labelling a triangle**
A capital letter is used to identify each vertex. A triangle with vertices A, B, and C is known as ▲ABC. The symbol "▲" can be used to represent the word triangle.

vertex the point where two sides meet

perimeter the length of the outside frame

side one of three sides

angle the amount of turn between two straight lines about a fixed point

base
side on which a triangle "rests"

Types of triangles

There are several types of triangle, each with specific features, or properties. A triangle is classified according to the length of its sides or the size of its angles.

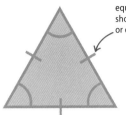

equal sides are shown by a dash or double dash

◁ **Equilateral triangle**
A triangle with three equal sides and three equal angles, each of which measures 60°.

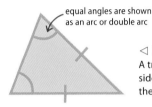

equal angles are shown as an arc or double arc

◁ **Isosceles triangle**
A triangle with two equal sides. The angles opposite these sides are also equal.

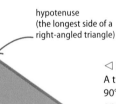

hypotenuse (the longest side of a right-angled triangle)

◁ **Right-angled triangle**
A triangle with an angle of 90° (a right angle). The side opposite the right angle is called the hypotenuse.

right angle

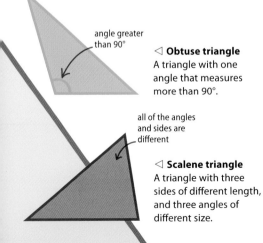

angle greater than 90°

◁ **Obtuse triangle**
A triangle with one angle that measures more than 90°.

all of the angles and sides are different

◁ **Scalene triangle**
A triangle with three sides of different length, and three angles of different size.

Interior angles of a triangle

A triangle has three interior angles at the points where each pair of sides meets. These angles always add up to 180°. If rearranged and placed together the angles make up a straight line, which always measures 180°.

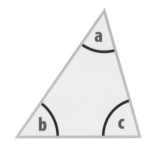

$$a + b + c = \mathbf{180°}$$

Proving that the sum of a triangle's angles is 180°

Adding a parallel line produces two types of relationships between angles that help prove that the interior sum of a triangle is 180°.

Draw a triangle, then add a line parallel to one side of the triangle, starting at its base, to create two new angles.

Corresponding angles are equal and alternate angles are equal; angles c, a, and b sit on a straight line so add up to 180°.

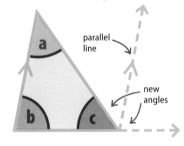

parallel line

new angles

alternate angles

corresponding angles

Exterior angles of a triangle

A triangle has three interior angles as well as three exterior angles. Exterior angles are found by extending each side of a triangle. The sum of the exterior angles of any triangle is 360°.

$$x + y + z = \mathbf{360°}$$

each exterior angle of a triangle is equal to the sum of the two opposite interior angles, so y = p + q

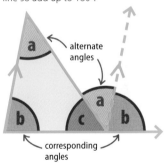

opposite interior angle (to y)

opposite interior angle (to y)

Constructing triangles

DRAWING (CONSTRUCTING) TRIANGLES REQUIRES A COMPASS, A RULER, AND A PROTRACTOR.

SEE ALSO

‹ 74-75 Tools in geometry
‹ 102-105 Constructions
‹ 106-107 Loci

To construct a triangle, not all the measurements for its sides and angles are required, as long as some of the measurements are known in the right combination.

What is needed?

A triangle can be constructed from just a few of its measurements, using a combination of the tools mentioned above, and its unknown measurements can be found from the result. A triangle can be constructed when the measurements of all three sides (SSS) are known, when two angles and the side in between are known (AAS), or when two sides and the angle between them are known (SAS). In addition, knowing either the SSS, the AAS, or the SAS measurements of two triangles will reveal whether they are the same size (congruent) – if the measurements are equal, the triangles are congruent.

Using triangles for 3-D graphics

3-D graphics are common in films, computer games, and the internet. What may be surprising is that they are created using triangles. An object is drawn as a series of basic shapes, which are then divided into triangles. When the shape of the triangles is changed, the object appears to move. Each triangle is coloured to bring the object to life.

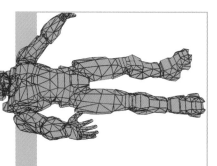

△ **Computer animation**
To create movement, a computer calculates the new shape of millions of shapes.

Constructing a triangle when three sides are known (SSS)

If the measurements of the three sides are given, for example, **5cm**, **4cm**, and **3cm**, it is possible to construct a triangle using a ruler and a compass, following the steps below.

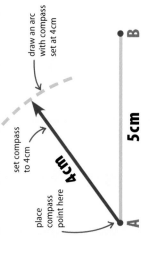

Draw the base line, using the longest length. Label the ends A and B. Set the compass to the second length, 4cm. Place the point of the compass on A and draw an arc.

place compass point here
set compass to 4cm
draw an arc with compass set at 4cm
5cm

Set the compass to the third length, 3cm. Place the point of the compass on B and draw another arc. Mark the spot where the arcs intersect (cross) as point C.

draw an arc with compass set at 3cm
point where two arcs cross is third point of triangle
set compass to 3cm
place point of compass here
5cm

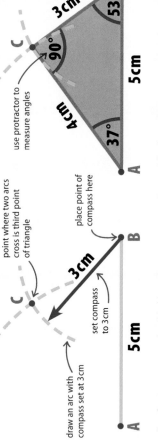

▷ **Join the points** to complete the triangle. Now use a protractor to find out the measurements of the angles. These will add up to 180° (90° + 53° + 37° = 180°).

use protractor to measure angles

Constructing a triangle when two angles and one side are known (AAS)

A triangle can be constructed when the two angles, for example, **73°** and **38°**, are given, along with the length of the side that falls between them, for example, **5cm**.

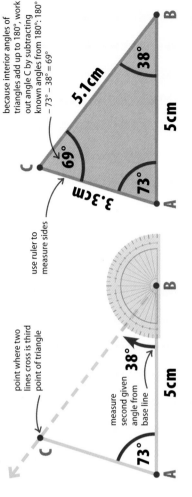

draw line from A at 73° angle

measure first given angle from base line

5cm

73°

Draw the base line of the triangle, here 5cm. Label the ends A and B. Place the protractor over A and measure the first angle, 73°. Draw a side of the triangle from A.

point where two lines cross is third point of triangle

measure second given angle from base line

73°   **38°**

5cm

Place the protractor over point B and mark 38°. Draw another side of the triangle from B. Point C is where the two new lines meet.

because interior angles of triangles add up to 180°, work out angle C by subtracting known angles from 180°: 180° − 73° − 38° = 69°

use ruler to measure sides

C   69°   5.1cm

73°   3.3cm   38°

A   5cm   B

Join the points to complete the triangle. Work out the unknown angle, and use a ruler to measure the two unknown sides.

Constructing a triangle when two sides and the angle in between are known (SAS)

Using the measurements for two of a triangle's sides, for example, **5cm** and **4.5cm**, and the angle between them, for example **50°**, it is possible to construct a triangle.

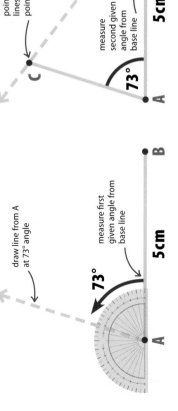

draw line at 50° from base line

measure given angle between two sides from base line

50°

5cm

A   B

Draw the base line, using the longest length. Label the ends A and B. Place the protractor over point A and mark 50°. Draw a line from A that runs through 50°. This line will be the next side of the triangle.

point C is where arc and line meet

C

4.5cm

50°

A   5cm   B

Set the compass to the second length, 4.5cm. Place the compass on point A and draw an arc. Point C is found when the arc intersects the line through point A.

measure unknown angles with protractor

measure unknown side with ruler

C   71°   4cm

4.5cm   59°

50°

A   5cm   B

Join the points to complete the triangle. Use a protractor to find the unknown angles and a ruler to find the length of the unknown side.

Congruent triangles

TRIANGLES THAT ARE EXACTLY THE SAME SHAPE AND SIZE.

SEE ALSO

❮ **90–91** Translations
❮ **92–93** Rotations
❮ **94–95** Reflections

Identical triangles

Two or more triangles are congruent if their sides are the same length and their corresponding interior angles are the same size. In addition to sides and angles, all other properties of congruent triangles are the same, for example, area. Like other shapes, congruent triangles can be translated, rotated, and reflected, so they may appear to look different, even if they remain the same size and have identical angles.

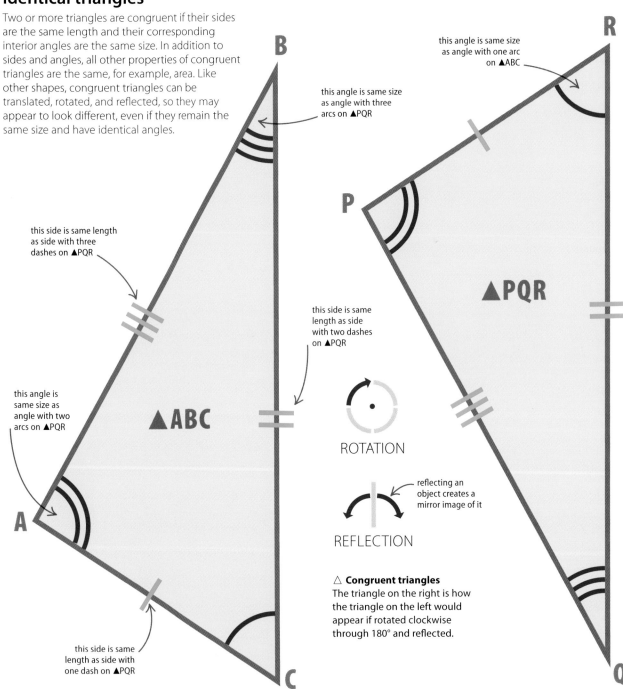

this angle is same size as angle with three arcs on ▲PQR

this angle is same size as angle with one arc on ▲ABC

this side is same length as side with three dashes on ▲PQR

this side is same length as side with two dashes on ▲PQR

▲PQR

this angle is same size as angle with two arcs on ▲PQR

▲ABC

ROTATION

reflecting an object creates a mirror image of it

REFLECTION

△ **Congruent triangles**
The triangle on the right is how the triangle on the left would appear if rotated clockwise through 180° and reflected.

this side is same length as side with one dash on ▲PQR

How to tell if triangles are congruent

It is possible to tell if two triangles are congruent without knowing the lengths of all of the sides or the sizes of all of the angles – knowing just three measurements will do. There are four groups of measurements.

▷ **Side, side, side (SSS)**
When all three sides of a triangle are the same as the corresponding three sides of another triangle, the two triangles are congruent.

▷ **Angle, angle, side (AAS)**
When two angles and any one side of a triangle are equal to two angles and the corresponding side of another triangle, the two triangles are congruent.

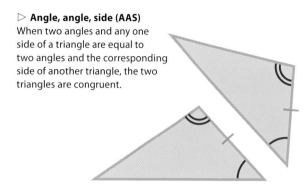

▷ **Side, angle, side (SAS)**
When two sides and the angle between them (called the included angle) of a triangle are equal to two sides and the included angle of another triangle, the two triangles are congruent.

▷ **Right angle, hypotenuse, side (RHS)**
When the hypotenuse and one other side of a right-angled triangle are equal to the hypotenuse and one side of another right-angled triangle, the two triangles are congruent.

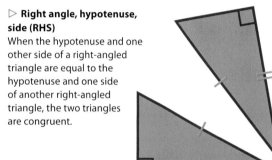

Proving an isosceles triangle has two equal angles

An isosceles triangle has two equal sides. Drawing a perpendicular line helps prove that it has two equal angles too.

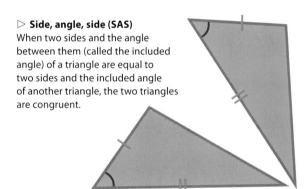

Draw a line perpendicular (at right angles) to the base of an isosceles triangle. This creates two new right-angled triangles. They are congruent – identical in every way.

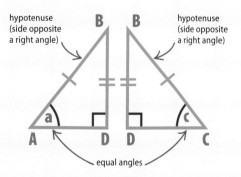

▷ **The perpendicular line** is common to both triangles. The two triangles have equal hypotenuses, another pair of equal sides, and right angles. The triangles are congruent (RHS) so angles "a" and "c" are equal.

Area of a triangle

AREA IS THE COMPLETE SPACE INSIDE A TRIANGLE.

SEE ALSO

⟨ **108–109** Triangles

Area of a circle **134–135** ⟩

Formulas **169–171** ⟩

What is area?

The area of a shape is the amount of space that fits inside its outline, or perimeter. It is measured in squared units, such as cm². If the length of the base and vertical height of a triangle are known, these values can be used to find the area of the triangle, using a simple formula, which is shown below.

$$\text{area} = \frac{1}{2} \times \text{base} \times \text{vertical height}$$

this is the formula for finding the area of a triangle

area is the space inside a triangle's frame

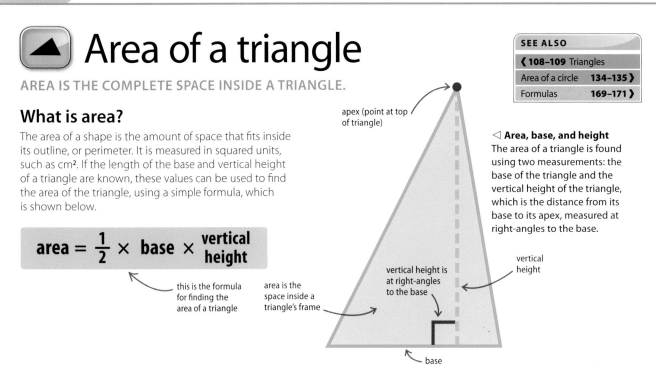

apex (point at top of triangle)

◁ **Area, base, and height**
The area of a triangle is found using two measurements: the base of the triangle and the vertical height of the triangle, which is the distance from its base to its apex, measured at right-angles to the base.

vertical height is at right-angles to the base

vertical height

base

Base and vertical height

Finding the area of a triangle requires two measurements: the base and the vertical height. The side on which a triangle "sits" is called the base. The vertical height is a line formed at right-angles to the base from the apex. Any one of the three sides of a triangle can act as the base in the area formula.

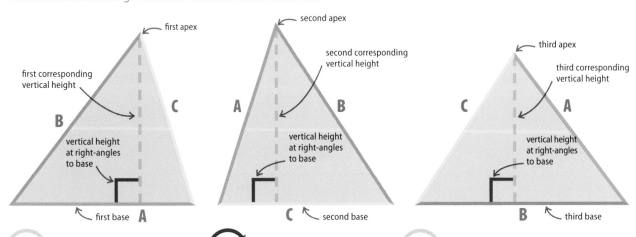

first apex

first corresponding vertical height

B

C

vertical height at right-angles to base

first base **A**

△ **First base**
The area of the triangle can be found using the orange side (A) as the "base" needed for the formula. The corresponding vertical height is the distance from the base of the triangle to its apex (highest point).

second apex

second corresponding vertical height

A

B

vertical height at right-angles to base

C second base

△ **Second base**
Any one of the triangle's three sides can act as its base. Here the triangle is rotated so that the green side (C) is its base. The corresponding vertical height is the distance from the base to the apex.

third apex

third corresponding vertical height

C

A

vertical height at right-angles to base

B third base

△ **Third base**
The triangle is rotated again, so that the purple side (B) is its base. The corresponding vertical height is the distance from the base to the apex. The area of the triangle is the same, whichever side is used as the base in the formula.

Finding the area of a triangle

To calculate the area of a triangle, substitute the given values for the base and vertical height into the formula. Then work through the multiplication shown by the formula (½ × base × vertical height).

▷ **An acute-angled triangle**
The base of this triangle is 6cm and its vertical height is 3cm. Find the area of the triangle using the formula.

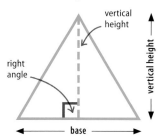

area is the space inside the triangle

base

vertical height

base = 6cm

vertical height = 3cm

First, write down the formula for the area of a triangle.

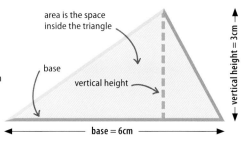

$$area = \frac{1}{2} \times base \times \text{vertical height}$$

Then, substitute the lengths that are known into the formula.

$$area = \frac{1}{2} \times 6 \times 3$$

Work through the multiplication in the formula to find the answer. In this example, ½ × 6 × 3 = 9. Add the units of area to the answer, here cm².

area is measured in squared units

$$area = 9cm^2$$

▷ **An obtuse-angled triangle**
The base of this triangle is 3cm and its vertical height is 4cm. Find the area of the triangle using the formula. The formula and the steps are the same for all types of triangles.

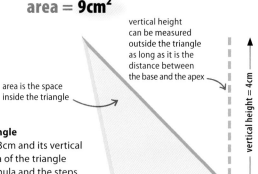

vertical height can be measured outside the triangle as long as it is the distance between the base and the apex

area is the space inside the triangle

vertical height = 4cm

base = 3cm

First, write down the formula for the area of a triangle.

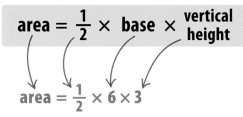

$$area = \frac{1}{2} \times base \times \text{vertical height}$$

Then, substitute the lengths that are known into the formula.

$$area = \frac{1}{2} \times 3 \times 4$$

Work through the multiplication to find the answer, and add the appropriate units of area.

½ × 3 × 4 = 6

area is measured in squared units

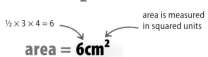

$$area = 6cm^2$$

Why the formula works

By adjusting the shape of a triangle, it can be converted into a rectangle. This process makes the formula for a triangle easier to understand.

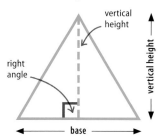

vertical height

right angle

vertical height

base

Draw any triangle and label its base and vertical height.

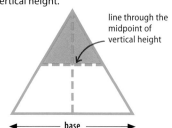

line through the midpoint of vertical height

base

Draw a line through the midpoint of the vertical height, which is parallel to the base.

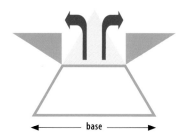

base

This creates two new triangles. These can be rotated around the triangle to form a rectangle. This has exactly the same area as the original triangle.

½ the vertical height of the triangle

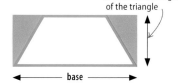

base

The original triangle's area is found using the formula for the area of a rectangle (b × h). Both shapes have the same base; the rectangle's height is ½ the height of the triangle. This gives the area of the triangle formula: ½ × base × vertical height.

Finding the base of a triangle using the area and height

The formula for the area of a triangle can also be used to find the length of the base, if the area and height are known. Given the area and height of the triangle, the formula needs to be rearranged to find the length of the triangle's base.

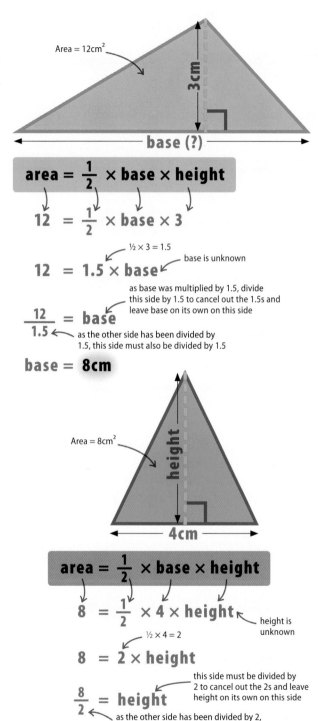

First, write down the formula for the area of a triangle. The formula states that the area of a triangle is equal to ½ multiplied by the length of the base, multiplied by the height.

$$\text{area} = \frac{1}{2} \times \text{base} \times \text{height}$$

Substitute the known values into the formula. Here the values of the area (12cm²) and the height (3cm) are known.

$$12 = \frac{1}{2} \times \text{base} \times 3$$

Simplify the formula as far as possible, by multiplying the ½ by the height. This answer is 1.5.

½ × 3 = 1.5
base is unknown

$$12 = 1.5 \times \text{base}$$

Make the base the subject of the formula by rearranging it. In this example both sides are divided by 1.5.

as base was multiplied by 1.5, divide this side by 1.5 to cancel out the 1.5s and leave base on its own on this side

$$\frac{12}{1.5} = \text{base}$$

as the other side has been divided by 1.5, this side must also be divided by 1.5

Work out the final answer by dividing 12 (area) by 1.5. In this example, the answer is 8cm.

$$\text{base} = \textbf{8cm}$$

Finding the vertical height of a triangle using the area and base

The formula for area of a triangle can also be used to find its height, if the area and base are known. Given the area and the length of the base of the triangle, the formula needs to be rearranged to find the height of the triangle.

First, write down the formula. This shows that the area of a triangle equals ½ multiplied by its base, multiplied by its height.

$$\text{area} = \frac{1}{2} \times \text{base} \times \text{height}$$

Substitute the known values into the formula. Here the values of the area (8cm²) and the base (4cm) are known.

$$8 = \frac{1}{2} \times 4 \times \text{height}$$

height is unknown

Simplify the equation as far as possible, by multiplying the ½ by the base. In this example, the answer is 2.

½ × 4 = 2

$$8 = 2 \times \text{height}$$

Make the height the subject of the formula by rearranging it. In this example both sides are divided by 2.

this side must be divided by 2 to cancel out the 2s and leave height on its own on this side

$$\frac{8}{2} = \text{height}$$

as the other side has been divided by 2, this side must also be divided by 2

Work out the final answer by dividing 8 (the area) by 2 (½ the base). In this example the answer is 4cm.

$$\text{height} = \textbf{4cm}$$

Similar triangles

TWO TRIANGLES THAT ARE EXACTLY THE SAME SHAPE BUT NOT
THE SAME SIZE ARE CALLED SIMILAR TRIANGLES.

SEE ALSO

❬ **48–51** Ratio and proportion
❬ **96–97** Enlargements
❬ **108–109** Triangles

What are similar triangles?

Similar triangles are made by making bigger or smaller copies of a triangle,
a transformation known as enlargement. Each of the triangles have equal
corresponding angles, and corresponding sides that are in proportion to
one another, for example each side of triangle ABC below is twice the
length of each side on triangle $A_2B_2C_2$. There are four different ways to check
if a pair of triangles are similar (see p.118), and if two triangles are known to
be similar, their properties can be used to find the lengths of missing sides.

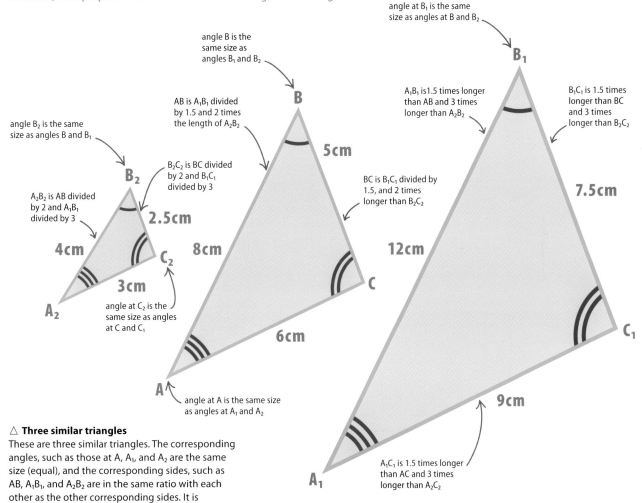

△ **Three similar triangles**
These are three similar triangles. The corresponding
angles, such as those at A, A_1, and A_2 are the same
size (equal), and the corresponding sides, such as
AB, A_1B_1, and A_2B_2 are in the same ratio with each
other as the other corresponding sides. It is
possible to check this by dividing each side of one
triangle by the corresponding side of another
triangle – if the answers are all equal, the sides are
in proportion to each other.

WHEN ARE TWO TRIANGLES SIMILAR?

It is possible to see if two triangles are similar without measuring every angle and every side. This can be done by looking at the following corresponding measurements for both triangles: two angles, all three sides, a pair of sides with an angle between them, or if the triangles are right-angled, the hypotenuse and another side.

Angle, angle AA

When two angles of one triangle are equal to two angles of another triangle then all the corresponding angles are equal in pairs, so the two triangles are similar.

$$U = U_1$$
$$V = V_1$$

angle $V_1 = V$

V_1

angle $V = V_1$

V

angle $U = U_1$

U_1

angle $U_1 = U$

W_1

U W

Side, angle, side (S) A (S)

When two triangles have two pairs of corresponding sides that are in the same ratio and the angles between these two sides are equal, the two triangles are similar.

$$\frac{PR}{P_1R_1} = \frac{PQ}{P_1Q_1} \quad \text{and} \quad P = P_1$$

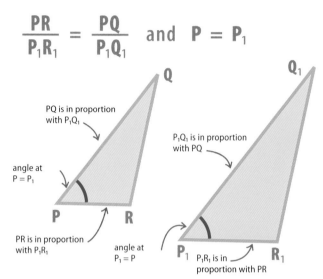

Q Q_1

PQ is in proportion with P_1Q_1

P_1Q_1 is in proportion with PQ

angle at $P = P_1$

P   R

PR is in proportion with P_1R_1

angle at $P_1 = P$

P_1 P_1R_1 is in proportion with PR R_1

Side, side, side (S) (S) (S)

When two triangles have three pairs of corresponding sides that are in the same ratio, then the two triangles are similar.

$$\frac{AB}{A_1B_1} = \frac{AC}{A_1C_1} = \frac{BC}{B_1C_1}$$

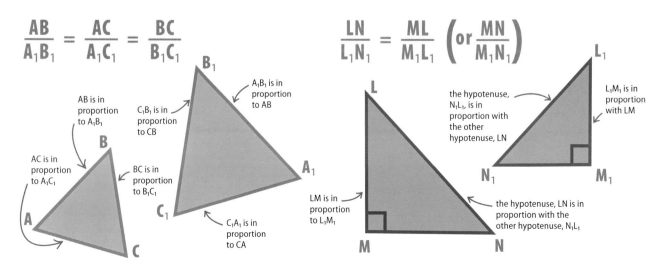

B_1

A_1B_1 is in proportion to AB

AB is in proportion to A_1B_1

C_1B_1 is in proportion to CB

B

AC is in proportion to A_1C_1

BC is in proportion to B_1C_1

A_1

A

C_1

C_1A_1 is in proportion to CA

C

Right-angle, hypotenuse, side R (H) (S)

If the ratio between the hypotenuses of two right-angled triangles is the same as the ratio between another pair of corresponding sides, then the two triangles are similar.

$$\frac{LN}{L_1N_1} = \frac{ML}{M_1L_1} \left(\text{or} \frac{MN}{M_1N_1} \right)$$

L_1

L

the hypotenuse, N_1L_1, is in proportion with the other hypotenuse, LN

L_1M_1 is in proportion with LM

N_1 M_1

LM is in proportion to L_1M_1

the hypotenuse, LN is in proportion with the other hypotenuse, N_1L_1

M   N

MISSING SIDES IN SIMILAR TRIANGLES

The proportional relationships between the sides of similar triangles can be used to find the value of sides that are missing, if the lengths of some of the sides are known.

▷ **Similar triangles**
Triangles ABC and ADE are similar (AA). The missing values of AD and BC can be found using the ratios between the known sides.

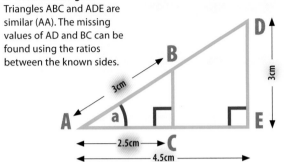

Finding the length of BC

To find the length of BC, use the ratio between BC and its corresponding side DE, and the ratio between a pair of sides where both the lengths are known – AE and AC.

Write out the ratios between the two pairs of sides, each with the longer side above the shorter side. These ratios are equal.

$$\frac{DE}{BC} = \frac{AE}{AC}$$

Substitute the values that are known into the ratios. The numbers can now be rearranged to find the length of BC.

$$\frac{3}{BC} = \frac{4.5}{2.5}$$

Rearrange the equation to isolate BC. This may take more than one step. First multiply both sides of the equation by BC.

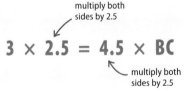

multiply both sides by BC

$$3 = \frac{4.5}{2.5} \times BC$$

Then rearrange the equation again. This time multipy both sides of the equation by 2.5.

multiply both sides by 2.5

$$3 \times 2.5 = 4.5 \times BC$$

multiply both sides by 2.5

BC can now be isolated by rearranging the equation one more time – divide both sides of the equation by 4.5.

divide both sides by 4.5

$$BC = \frac{3 \times 2.5}{4.5}$$

divide both sides by 4.5

Do the multiplication to find the answer, add the units, and round to a sensible number of decimal places.

1.6666.... is rounded to 2 decimal places

$$BC = \mathbf{1.67cm}$$

Finding the length of AD

To find the length of AD, use the ratio between AD and its corresponding side AB, and the ratio between a pair of sides where both the lengths are known – AE and AC.

Write out the ratios between the two pairs of sides, each with the longer side above the shorter side. These ratios are equal.

$$\frac{AD}{AB} = \frac{AE}{AC}$$

Substitute the values that are known into the ratios. The numbers can now be rearranged to find the length of AD.

AD is the unknown

$$\frac{AD}{3} = \frac{4.5}{2.5}$$

Rearrange the equation to isolate AD. In this example this is done by multiplying both sides of the equation by 3.

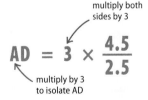

multiply both sides by 3

$$AD = 3 \times \frac{4.5}{2.5}$$

multiply by 3 to isolate AD

Do the multiplication to find the answer, and add the units to the answer that has been found. This is the length of AD.

$$AD = \mathbf{5.4cm}$$

Pythagoras' theorem

PYTHAGORAS' THEOREM IS USED TO FIND THE LENGTH OF
MISSING SIDES IN RIGHT-ANGLED TRIANGLES.

SEE ALSO
❮ **32–35** Powers and roots
❮ **108–109** Triangles
❮ **114–116** Area of a triangle
Formulas **169–171** ❯

If the lengths of two sides of a right-angled triangle are known, the
length of the third side can be worked out using Pythagoras' theorem.

What is Pythagoras' theorem?

The basic principle of Pythagoras' theorem is that
squaring the two smaller sides of a right-angled
triangle (i.e. multiplying each side by itself) and
adding the results together will equal the square
of the longest side. The idea of "squaring" each side
can be shown literally as three different square
shapes. On the right, a square on each of the three
sides shows how the biggest square has the same
area as the other two squares put together.

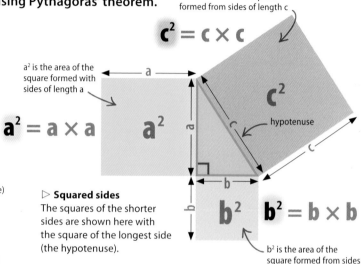

c^2 is the area of the square
formed from sides of length c

$$c^2 = c \times c$$

a^2 is the area of the
square formed with
sides of length a

$$a^2 = a \times a$$

hypotenuse

$$b^2 = b \times b$$

b^2 is the area of the
square formed from sides
of length b

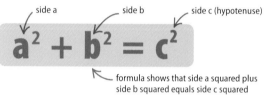

side a   side b   side c (hypotenuse)

$$a^2 + b^2 = c^2$$

formula shows that side a squared plus
side b squared equals side c squared

▷ **Squared sides**
The squares of the shorter
sides are shown here with
the square of the longest side
(the hypotenuse).

If the formula is used with values substituted for the sides a, b, and c,
Pythagoras' theorem can be shown to be true. Here the length of c
(the hypotenuse) is 5, while the lengths of a and b are 4 and 3.

a is 4   b is 3   c is 5

$$a^2 + b^2 = c^2$$

$$4^2 + 3^2 = 5^2$$

4×4 3×3 5×5

$$16 + 9 = 25$$

squares of two
shorter sides
added

square of
hypotenuse

△ **Pythagoras in action**
In the equation the squares of the two
shorter sides (4 and 3) added together
equal the square of the hypotenuse (5),
proving that Pythagoras' theorem works.

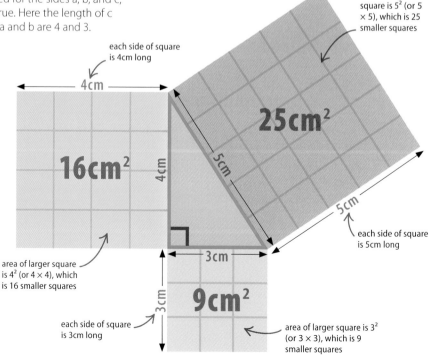

each side of square
is 4cm long

4cm

16cm²

4cm

5cm

25cm²

5cm

each side of square
is 5cm long

area of larger
square is 5² (or 5
× 5), which is 25
smaller squares

3cm

area of larger square
is 4² (or 4 × 4), which
is 16 smaller squares

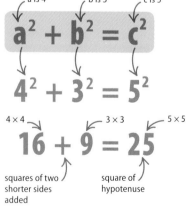

3cm

each side of square
is 3cm long

9cm²

area of larger square is 3²
(or 3 × 3), which is 9
smaller squares

Find the value of the hypotenuse

Pythagoras' theorem can be used to find the value of the length of the longest side (the hypotenuse) in a right-angled triangle when the lengths of the two shorter sides are known. This example shows how this works, if the two known sides are 3.5cm and 7.2cm in length.

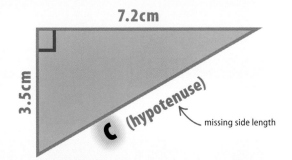

7.2cm

3.5cm

c (hypotenuse)

missing side length

$$a^2 + b^2 = c^2$$

First, take the formula for Pythagoras' theorem.

one side   other side   hypotenuse missing

$$3.5^2 + 7.2^2 = c^2$$

Substitute the values given into the formula, in this example, 3.5 and 7.2.

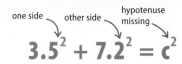

3.5 × 3.5 equals   7.2 × 7.2 equals

$$12.25 + 51.84 = c^2$$

Work out the squares of each of the triangle's known sides by multiplying them.

12.25 + 51.84 equals

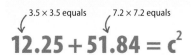

$$64.09 = c^2$$

Add these answers together to find the square of the hypotenuse.

sign means square root

$$\sqrt{64.09} = \sqrt{c^2}$$

the square root of 64.09 is the same as the square root of c^2

Use a calculator to find the square root of 64.09. This gives the length of side c.

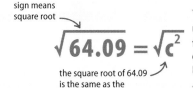

answer given to two decimal places

The square root is the length of the hypotenuse.

$$c = \textbf{8.01cm}$$

Find the value of another side

The theorem can be rearranged to find the length of either of the two sides of a right-angled triangle that are not the hypotenuse. The length of the hypotenuse and one other side must be known. This example shows how this works with a side of 5cm and a hypotenuse of 13cm.

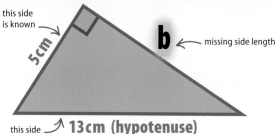

this side is known

b

missing side length

5cm

this side is known   **13cm (hypotenuse)**

$$a^2 + b^2 = c^2$$

To work out the length of side b, first take the formula for Pythagoras' theorem.

known side   hypotenuse

$$5^2 + b^2 = 13^2$$

unknown side

Substitute the values given into the formula. In this example, 5 and 13.

unknown side is now result of equation

$$13^2 - 5^2 = b^2$$

hypotenuse now at start of formula

Rearrange the equation by subtracting 5^2 from each side. This isolates b^2 on one side because $5^2 - 5^2$ cancels out.

$$169 - 25 = b^2$$

13 × 13 equals   5 × 5 equals

Work out the squares of the two known sides of the triangle.

$$144 = b^2$$

Subtract these squares to find the square of the unknown side.

sign means square root

$$\sqrt{144} = \sqrt{b^2}$$

the square root of 144 is the same as the square root of b^2

Find the square root of 144 for the length of the unknown side.

length of missing side

The square root is the length of side b.

$$b = \textbf{12cm}$$

Quadrilaterals

A QUADRILATERAL IS A FOUR-SIDED POLYGON.
"QUAD" MEANS FOUR AND "LATERAL" MEANS SIDE.

SEE ALSO
❮ 76–77 Angles
❮ 78–79 Straight lines
Polygons 126–129 ❯

Introducing quadrilaterals

A quadrilateral is a two-dimensional shape with four straight sides, four vertices (points where the sides meet), and four interior angles. The interior angles of a quadrilateral always add up to 360°. An exterior angle and its corresponding interior angle always add up to 180° because they form a straight line. There are several types of quadrilaterals, each with different properties.

vertex, one of four vertices

one of four sides

diagonal

one of four interior angles

interior angle

line extends to form exterior angle

interior and exterior angle add up to 180°

△ **Interior angles**
If a single diagonal line is drawn from any one corner to the opposite corner, the quadrilateral is divided into two triangles. The sum the interior angles of any triangle is 180°, so the sum of the interior angles of a quadrilateral is 2 × 180°.

▽ **Types of quadrilaterals**
Each type of quadrilateral is grouped and named according to its properties. There are regular and irregular quadrilaterals. A regular quadrilateral has equal sides and angles, whereas an irregular quadrilateral has sides and angles of different sizes.

START

? Are all the interior angles right-angles?
| YES | NO |

? Are all the sides the same length?
| YES | NO |

? Are the opposite angles equal?
| YES | NO |

? Are two of the sides parallel?
| YES | NO |

? Are all the sides the same length?
| YES | NO |

? Are there adjacent sides of the same length?
| YES | NO |

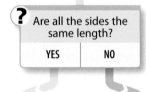
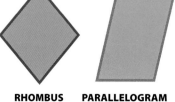
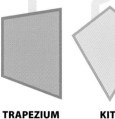

SQUARE   **RECTANGLE**   **RHOMBUS**   **PARALLELOGRAM**   **TRAPEZIUM**   **KITE**   **IRREGULAR**

PROPERTIES OF QUADRILATERALS

Each type of quadrilateral has its own name and a number of unique properties. Knowing just some of the properties of a shape can help distinguish one type of quadrilateral from another. Six of the more common quadrilaterals are shown below with their respective properties.

Square

A square has four equal angles (right angles) and four sides of equal length. The opposite sides of a square are parallel. The diagonals bisect – cut into two equal parts – each other at 90° (right angles).

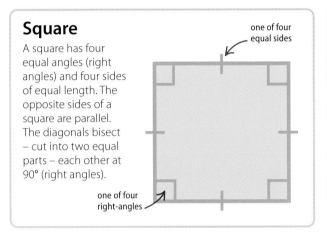

one of four equal sides

one of four right-angles

Rectangle

A rectangle has four right angles and two pairs of opposite sides of equal length. Adjacent sides are not of equal length. The opposite sides are parallel and the diagonals bisect each other.

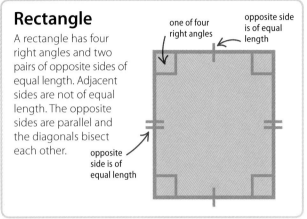

one of four right angles

opposite side is of equal length

opposite side is of equal length

Rhombus

All sides of a rhombus are of equal length. The opposite angles are equal and the opposite sides are parallel. The diagonals bisect each other at right angles.

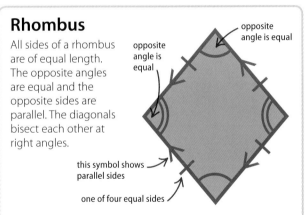

opposite angle is equal

opposite angle is equal

this symbol shows parallel sides

one of four equal sides

Parallelogram

The opposite sides of a parallelogram are parallel and are of equal length. Adjacent sides are not of equal length. The opposite angles are equal and the diagonals bisect each other in the centre of the shape.

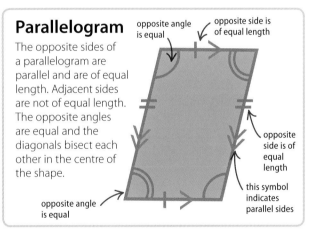

opposite angle is equal

opposite side is of equal length

opposite side is of equal length

this symbol indicates parallel sides

opposite angle is equal

Trapezium

A trapezium, also known as a trapezoid, has one pair of opposite sides that are parallel. These sides are not equal in length.

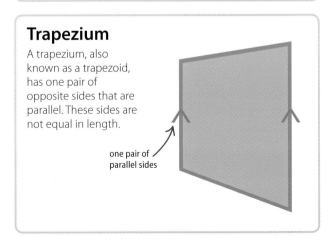

one pair of parallel sides

Kite

A kite has two pairs of adjacent sides that are equal in length. Opposite sides are not of equal length. It has one pair of opposite angles that are equal and another pair of angles of different values.

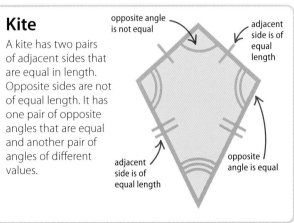

opposite angle is not equal

adjacent side is of equal length

adjacent side is of equal length

opposite angle is equal

FINDING THE AREA OF QUADRILATERALS

Area is the space inside the frame of a two-dimensional shape. Area is measured in square units, for example, cm^2. Formulas are used to calculate the areas of many types of shapes. Each type of quadrilateral has a unique formula for calculating its area.

Finding the area of a square

The area of a square is found by multiplying its length by its width. As its length and width are equal in size, the formula is the square of a side.

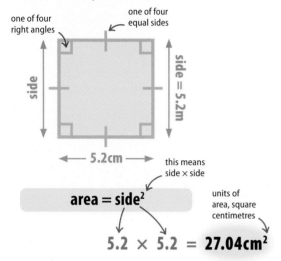

one of four right angles

one of four equal sides

side

side = 5.2m

5.2cm

this means side × side

$$\text{area} = \text{side}^2$$

units of area, square centimetres

$$5.2 \times 5.2 = 27.04cm^2$$

△ **Multiply sides**
In this example, each of the four sides measures 5.2cm. To find the area of this square, multiply 5.2 by 5.2.

Finding the area of a rectangle

The area of a rectangle is found by multiplying its base by its height.

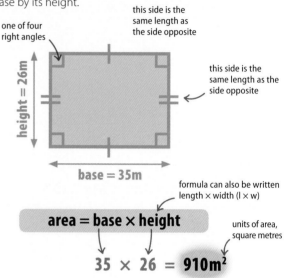

this side is the same length as the side opposite

one of four right angles

height = 26m

this side is the same length as the side opposite

base = 35m

formula can also be written length × width (l × w)

$$\text{area} = \text{base} \times \text{height}$$

units of area, square metres

$$35 \times 26 = 910m^2$$

△ **Multiply base by height**
The height (or width) of this rectangle is 26m, and its base (or length) measures 35m. Multiply these two measurements together to find the area.

Finding the area of a rhombus

The area of a rhombus is found by multiplying the length of its base by its vertical height. The vertical height, also known as the perpendicular height, is the vertical distance from the top (vertex) of a shape to the base opposite. The vertical height is at right angles to the base.

▷ **Vertical height**
Finding the area of a rhombus depends on knowing its vertical height. In this example, the vertical height measures 8cm and its base is 9cm.

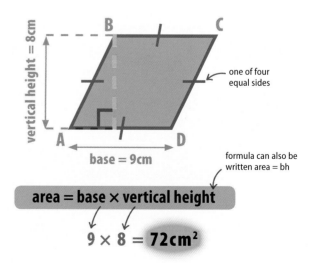

vertical height = 8cm

B   C

one of four equal sides

A   D

base = 9cm

formula can also be written area = bh

$$\text{area} = \text{base} \times \text{vertical height}$$

$$9 \times 8 = 72cm^2$$

Finding the area of a parallelogram

Like the area of a rhombus, the area of a parallelogram is found by multiplying the length of its base by its vertical height.

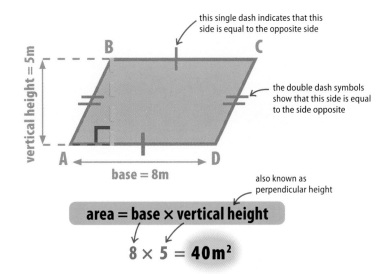

this single dash indicates that this side is equal to the opposite side

the double dash symbols show that this side is equal to the side opposite

vertical height = 5m

base = 8m

also known as perpendicular height

▷ **Multiply base by vertical height**
It is important to remember that the slanted side, AB, is not the vertical height. This formula only works if the vertical height is used.

area = base × vertical height

$$8 \times 5 = 40m^2$$

Proving the opposite angles of a rhombus are equal

Creating two pairs of isosceles triangles by dividing a rhombus along two diagonals helps prove that the opposite angles of a rhombus are equal. An isosceles triangle has two equal sides and two equal angles.

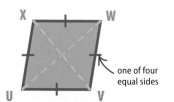

one of four equal sides

All the sides of a rhombus are equal in length. To show this a dash is used on each side.

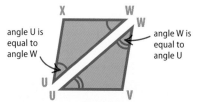

angle U is equal to angle W

angle W is equal to angle U

▷ **Divide the rhombus** along the diagonals to create two isosceles triangles. Each triangle has a pair of equal angles.

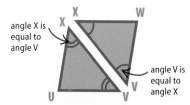

angle X is equal to angle V

angle V is equal to angle X

▷ **Dividing along the other diagonal** creates another pair of isosceles triangles.

Proving the opposite sides of a parallelogram are parallel

Creating a pair of congruent triangles by dividing a parallelogram along two diagonals helps prove that the opposite sides of a parallelogram are parallel. Congruent triangles are the same size and shape.

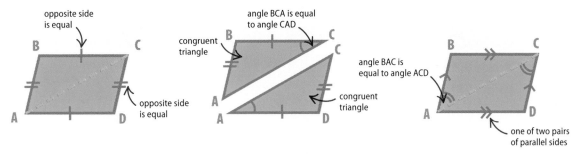

opposite side is equal

opposite side is equal

Opposites sides of a parallelogram are equal in length. To show this a dash and a double dash are used.

congruent triangle

angle BCA is equal to angle CAD

congruent triangle

▷ **The triangles ABC and ADC** are congruent. Angle BCA = CAD, and as these are alternate angles, BC is parallel to AD.

angle BAC is equal to ACD

one of two pairs of parallel sides

▷ **The triangles are congruent**, so angle BAC = ACD; as these are alternate angles, DC is parallel to AB.

Polygons

A CLOSED TWO-DIMENSIONAL SHAPE OF THREE OR MORE SIDES.

SEE ALSO
❰ 76–77 Angles
❰ 108–109 Triangles
❰ 112–113 Congruent triangles
❰ 122–125 Quadrilaterals

Polygons range from simple three-sided triangles and four-sided squares to more complicated shapes such as trapezoids and dodecagons. Polygons are named according to the number of sides and angles they have.

What is a polygon?

A polygon is a closed two-dimensional shape formed by straight lines that connect end to end at a point called a vertex. The interior angles of a polygon are usually smaller than the exterior angles, although the reverse is possible. Polygons with an interior angle of more than 180° are called re-entrant.

▷ **Parts of a polygon**
Regardless of shape, all polygons are made up the same parts – sides, vertices (connecting points), and interior and exterior angles.

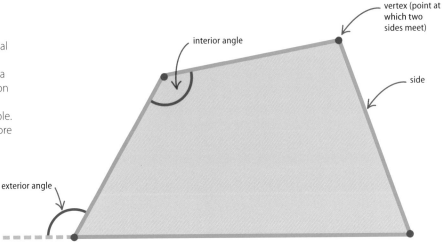

vertex (point at which two sides meet)

interior angle

side

exterior angle

Describing polygons

There are several ways to describe polygons. One is by the regularity or irregularity of their sides and angles. A polygon is regular when all of its sides and angles are equal. An irregular polygon has at least two sides or two angles that are different.

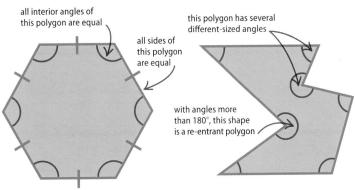

all interior angles of this polygon are equal

all sides of this polygon are equal

this polygon has several different-sized angles

with angles more than 180°, this shape is a re-entrant polygon

△ **Regular**
All the sides and all the angles of regular polygons are equal. This hexagon has six equal sides and six equal angles, making it regular.

△ **Irregular**
In an irregular polygon, all the sides and angles are not the same. This heptagon has many different-sized angles, making it irregular.

Equal angles or equal sides?

All the angles and all the sides of a regular polygon are equal – in other words, the polygon is both equiangular and equilateral. In certain polygons, only the angles (equiangular) or only the sides (equilateral) are equal.

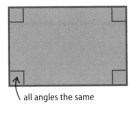

◁ **Equiangular**
A rectangle is an equiangular quadrilateral. Its angles are all equal, but not all its sides are equal.

all angles the same

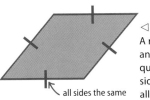

◁ **Equilateral**
A rhombus is an equilateral quadrilateral. All its sides are equal, but all its angles are not.

all sides the same

Naming polygons

Regardless of whether a polygon is regular or irregular, the number of sides it has always equals the number of its angles. This number is used in naming both kinds of polygons. For example, a polygon with six sides and angles is called a hexagon because "hex" is the prefix used to mean six. If all of its sides and angles are equal, it is known as a regular hexagon; if not, it is called an irregular hexagon.

Triangle

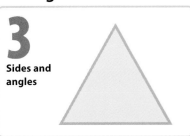

3
Sides and angles

Quadrilateral

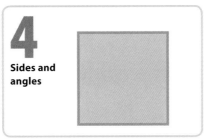

4
Sides and angles

Pentagon

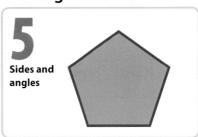

5
Sides and angles

Hexagon

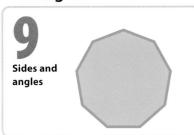

6
Sides and angles

Heptagon

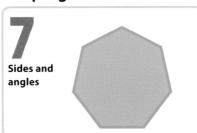

7
Sides and angles

Octagon

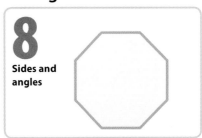

8
Sides and angles

Nonagon

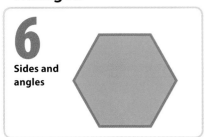

9
Sides and angles

Decagon

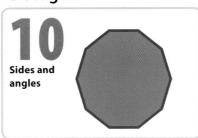

10
Sides and angles

Hendecagon

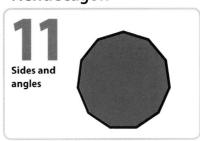

11
Sides and angles

Dodecagon

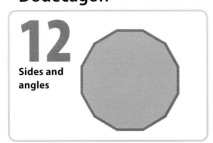

12
Sides and angles

Pentadecagon

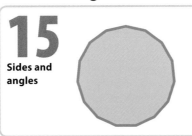

15
Sides and angles

Icosagon

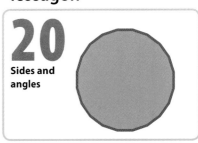

20
Sides and angles

PROPERTIES OF A POLYGON

There are an unlimited number of different polygons that can be drawn using straight lines. However, they all share some important properties.

Convex or concave

Regardless of how many angles a polygon has, it can be classified as either concave or convex. This difference is based on whether a polygon's interior angles are over 180° or not. A convex polygon can be easily identified because at least one its angles is over 180°.

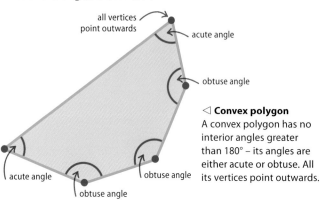

all vertices point outwards

acute angle

obtuse angle

◁ **Convex polygon**
A convex polygon has no interior angles greater than 180° – its angles are either acute or obtuse. All its vertices point outwards.

acute angle

obtuse angle

obtuse angle

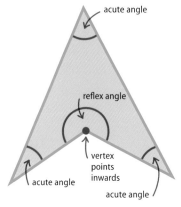

acute angle

reflex angle

vertex points inwards

acute angle

acute angle

◁ **Concave polygon**
At least one angle of a concave polygon is over 180°. This type of angle is known as a reflex angle. The vertex of the reflex angle points inwards, towards the centre of the shape.

Interior angle sum of polygons

The sum of the interior angles of both regular and irregular convex polygon depends on the number of sides the polygon has. The sum of the angles can be worked out by dividing the polygon into triangles.

convex quadrilateral

diagonal splits shape into two triangles

This quadrilateral is convex – all of its angles are smaller than 180°. The sum of its interior angles can be found easily, by breaking the shape down into triangles. This can be done by drawing in a diagonal line that connects two vertices that are not next to one another.

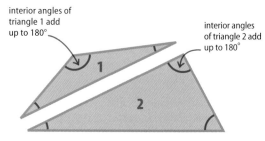

interior angles of triangle 1 add up to 180°

interior angles of triangle 2 add up to 180°

1

2

A quadrilateral can be split into two triangles. The sum of the angles of each triangle is 180°, so the sum of the angles of the quadrilateral is the sum of the angles of the two triangles added together: 2 × 180° = 360°.

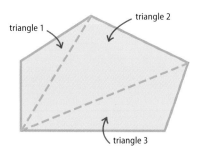

triangle 2

triangle 1

triangle 3

◁ **Irregular pentagon**
This pentagon can be split up into three triangles. The sum of its interior angles is the sum of the angles of the three triangles: 3 × 180° = 540°.

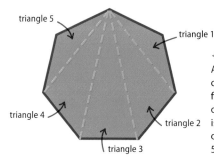

triangle 5

triangle 1

triangle 4

triangle 2

triangle 3

◁ **Regular heptagon**
A heptagon (7 sides) can be split up into five triangles. The sum of its interior angles is the sum of the angles of the five triangles: 5 × 180° = 900°.

A formula for the interior angle sum

The number of triangles a convex polygon can be split up into is always 2 fewer than the number of its sides. This means that a formula can be used to find the sum of the interior angles of any convex polygon.

Sum of interior angles = (n – 2) × 180°

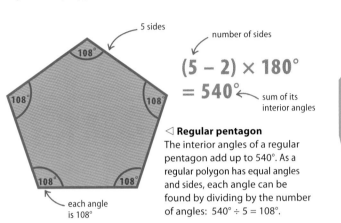

number of sides

$$(5 - 2) \times 180°$$
$$= 540°$$

sum of its interior angles

◁ **Regular pentagon**
The interior angles of a regular pentagon add up to 540°. As a regular polygon has equal angles and sides, each angle can be found by dividing by the number of angles: 540° ÷ 5 = 108°.

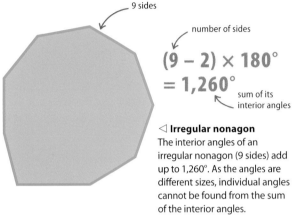

9 sides

number of sides

$$(9 - 2) \times 180°$$
$$= 1,260°$$

sum of its interior angles

◁ **Irregular nonagon**
The interior angles of an irregular nonagon (9 sides) add up to 1,260°. As the angles are different sizes, individual angles cannot be found from the sum of the interior angles.

Sum of exterior angles of a polygon

Imagine walking along the exterior of a polygon. Start at one vertex, and facing the next, walk towards it. At the next vertex, turn the number of degrees of the exterior angle until facing the following vertex, and repeat until you have been around all the vertices. In walking around the polygon, you will have turned a complete circle, or 360°. The exterior angles of any polygon always add up to 360°.

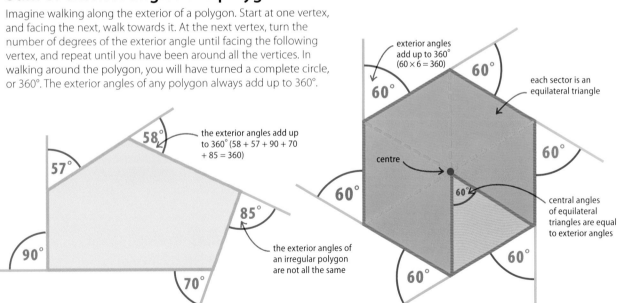

the exterior angles add up to 360° (58 + 57 + 90 + 70 + 85 = 360)

the exterior angles of an irregular polygon are not all the same

exterior angles add up to 360° (60 × 6 = 360)

each sector is an equilateral triangle

centre

central angles of equilateral triangles are equal to exterior angles

△ **Irregular pentagon**
The exterior angles of a polygon, regardless of whether it is regular or irregular add up to 360°. Another way to think about this is that, added together, the exterior angles of a polygon would form a complete circle.

△ **Regular hexagon**
The size of the exterior angles of a regular polygon can be found by dividing 360° by the number of sides the polygon has. A regular hexagon's central angles (formed by splitting the shape into 6 equilateral triangles) are the same as the exterior angles.

 Circles

A CIRCLE IS A CURVED LINE SURROUNDING A CENTRE POINT. EVERY POINT OF THIS CURVED LINE IS OF EQUAL DISTANCE FROM THE CENTRE POINT.

SEE ALSO

❰ 74–75 Tools in geometry

Circumference of a circle 132–133 ❱

Area of a circle 134–135 ❱

Properties of a circle

A circle can be folded into two identical halves, which means that it possesses "reflective symmetry" (see p.80). The line of this fold is one of the most important parts of a circle – its diameter. A circle may also be rotated about its centre and still fit into its own outline, giving it a "rotational symmetry" about its centre point.

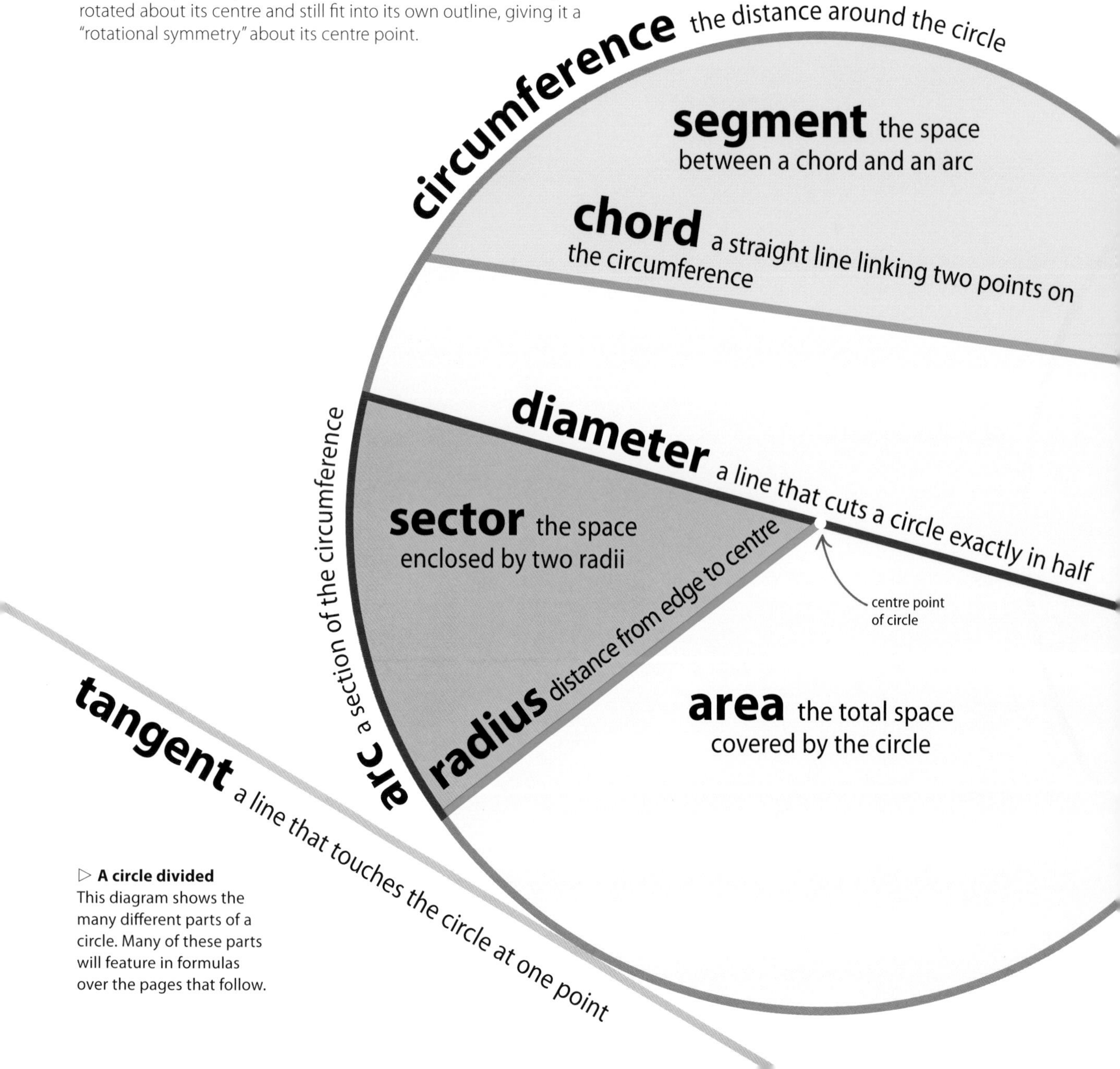

circumference the distance around the circle

segment the space between a chord and an arc

chord a straight line linking two points on the circumference

diameter a line that cuts a circle exactly in half

sector the space enclosed by two radii

centre point of circle

radius distance from edge to centre

area the total space covered by the circle

arc a section of the circumference

tangent a line that touches the circle at one point

▷ **A circle divided**
This diagram shows the many different parts of a circle. Many of these parts will feature in formulas over the pages that follow.

Parts of a circle

A circle can be measured and divided in various ways. Each of these has a specific name and character, and they are all shown below.

Radius
Any straight line from the centre of a circle to its circumference. The plural of radius is radii.

Diameter
Any straight line that passes through the centre from one side of a circle to the other.

Chord
Any straight line linking two points on a circle's circumference, but not passing through its centre.

Segment
The smaller of the two parts of a circle created when divided by a chord.

Circumference
The total length of the outside edge (perimeter) of a circle.

Arc
Any section of the circumference of a circle.

Sector
A "slice" of a circle, similar to the slice of a pie. It is enclosed by two radii and an arc.

Area
The amount of space inside a circle's circumference.

Tangent
A straight line that touches the circle at a single point.

How to draw a circle

Two instruments are needed to draw a circle – a compass and a pencil. The point of the compass marks the centre of the circle and the distance between the point and the pencil attached to the compass forms the circle's radius. A ruler is needed to measure the radius of the circle correctly.

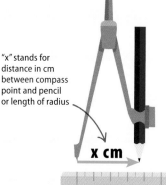

"x" stands for distance in cm between compass point and pencil or length of radius

x cm

use ruler to set length of radius

Set the compass. First, decide what the radius of the circle is, and then use a ruler to set the compass at this distance.

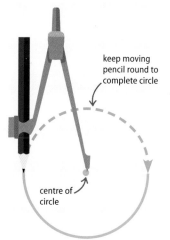

keep moving pencil round to complete circle

centre of circle

Decide where the centre of the circle is and then hold the point of the compass firmly in this place. Then put the pencil to the paper and move the pencil round to draw the circumference of the circle.

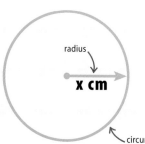

radius

x cm

circumference

The completed circle has a radius that is the same length as the distance that the compass was originally set to.

Circumference and diameter

THE DISTANCE AROUND THE EDGE OF A CIRCLE IS CALLED THE CIRCUMFERENCE; THE DISTANCE ACROSS THE MIDDLE IS THE DIAMETER.

SEE ALSO

❮ **48–51** Ratio and proportion
❮ **96–97** Enlargements
❮ **130–131** Circles
Area of a circle **134–135** ❯

All circles are similar because they have exactly the same shape. This means that all their measurements, including the circumference and the diameter, are in proportion to each other.

The number pi

The ratio between the circumference and diameter of a circle is a number called pi, which is written π. This number is used in many of the formulas associated with circles, including the formulas for the circumference and diameter.

symbol for pi

$$\pi = 3.14$$

value to 2 decimal places

◁ **The value of pi**
The numbers after the decimal point in pi go on for ever and in an unpredictable way. It starts 3.1415926 but is usually given to two decimal places.

Circumference (c)

The circumference is the distance around the edge of a circle. A circle's circumference can be found using the diameter or radius and the number pi. The diameter is always twice the length of the radius.

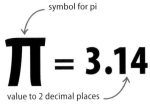

circumference   π is a constant   radius

$$c = 2\,\pi\,r$$

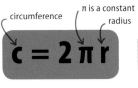

circumference   π is a constant   diameter

$$c = \pi\,d$$

◁ **Formulas**
There are two circumference formulas. One uses diameter and the other uses radius.

The formula for circumference shows that the circumference is equal to pi multiplied by the diameter of the circle.

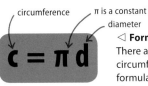

$$c = \pi\,d$$

d is the same as 2 × r, the formula can also be written C=2πr

Substitute known values into the formula for circumference. Here, the radius of the circle is known to be 3cm.

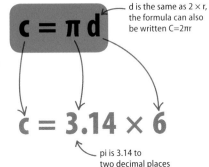

$$c = 3.14 \times 6$$

pi is 3.14 to two decimal places

Multiply the numbers to find the length of the circumference. Round the answer to a suitable number of decimal places.

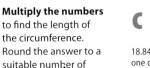

$$c = 18.8\text{cm}$$

18.84 is rounded to one decimal place

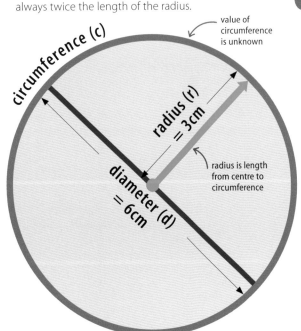

circumference (c)

value of circumference is unknown

radius (r) = 3cm

radius is length from centre to circumference

diameter (d) = 6cm

△ **Finding the circumference**
The length of a circle's circumference can be found if the length of the diameter is known, in this example the diameter is 6cm long.

Diameter (d)

The diameter is the distance across the middle of a circle. It is twice the length of the radius. A circle's diameter can be found by doubling the length of its radius, or by using its circumference and the number pi in the formula shown below. The formula is a rearranged version of the formula for the circumference of a circle.

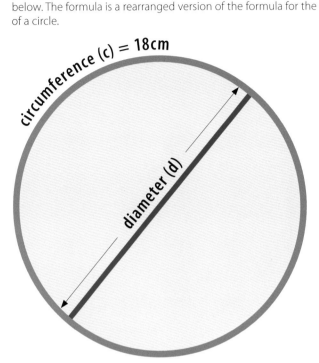

circumference (c) = 18cm

diameter (d)

△ **Finding the diameter**
This circle has a circumference of 18cm. Its diameter can be found using the formula given above.

diameter   circumference

The formula for diameter shows that the length of the diameter is equal to the length of the circumference divided by the number pi.

$$d = \frac{c}{\pi}$$

π is a constant

▼

Substitute known values into the formula for diameter. In the example shown here, the circumference of the circle is 18cm.

$$d = \frac{18}{\pi}$$

▼

Divide the circumference by the value of pi, 3.14, to find the length of the diameter.

$$d = \frac{18}{3.14}$$

more accurate to use π button on a calculator

▼

Round the answer to a suitable number of decimal places. In this example, the answer is given to two decimal places.

$$d = 5.73\text{cm}$$

the answer is given to two decimal places

LOOKING CLOSER

Why π ?

All circles are similar to one another. This means that corresponding lengths in circles, such as their diameters and circumferences, are always in proportion to each other. The number π is found by dividing the circumference of a circle by its diameter – any circle's circumference divided by its diameter always equals π – it is a constant value.

▷ **Similar circles**
As all circles are enlargements of each other, their diameters (d1, d2) and circumferences (c1, c2) are always in proportion to one another.

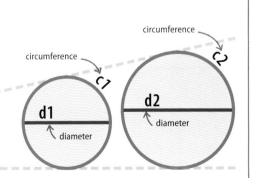

circumference

circumference

c1

c2

d1

d2

diameter

diameter

0•

Area of a circle

THE AREA OF A CIRCLE IS THE AMOUNT OF SPACE ENCLOSED INSIDE ITS PERIMETER (CIRCUMFERENCE).

SEE ALSO

❰ **130–131** Circles
❰ **132–133** Circumference and diameter
Formulas **169–171** ❱

The area of a circle can be found by using the measurements of either the radius or the diameter of the circle.

Finding the area of a circle

The area of a circle is measured in square units. It can be found using the radius of a circle (r) and the formula shown below. If the diameter is known but the radius is not, the radius can be found by dividing the diameter by 2.

In the formula for the area of a circle, πr² means π (pi) × radius × radius.

area of a circle → π is a fixed value
→ radius

$$\text{area} = \pi \, r^2$$

Substitute the known values into the formula, in this example the radius is 4cm.

$$\text{area} = 3.14 \times 4^2$$

π is 3.14 to 3 significant figures; a more accurate value can be found on a calculator

this means 4 × 4

Multiply the radius by itself as shown – this makes the last multiplication simpler.

$$\text{area} = 3.14 \times 16$$

4 × 4 = 16

Make sure the answer is in the right units (cm² here) and round it to a suitable number.

$$\text{area} = 50.24\text{cm}^2$$

answer is exactly 50.24

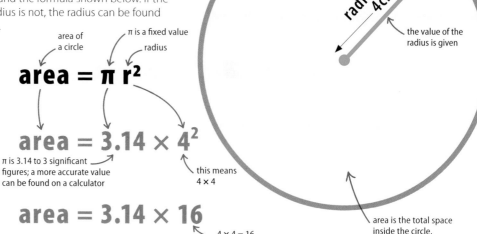

edge of circle is circumference

radius (r) = 4cm

the value of the radius is given

area is the total space inside the circle, shown in yellow

LOOKING CLOSER

Why does the formula for the area of a circle work?

The formula for the area of a circle can be proved by dividing a circle into segments, and rearranging the segments into a rectangular shape. The formula for the area of a rectangle (height × width) is simpler than that of the area for a circle. The rectangular shape's height is simply the length of a circle segment, which is the same as the radius of the circle. The width of the rectangular shape is half of the total segments, equivalent to half the circumference of the circle.

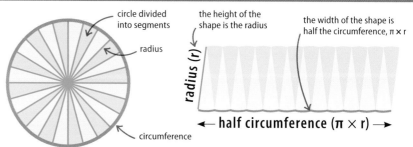

circle divided into segments

radius

circumference

the height of the shape is the radius

the width of the shape is half the circumference, π × r

radius (r)

← half circumference (π × r) →

Split any circle up into equal segments, making them as small as possible.

Lay the segments out in a rectangular shape. The area of a rectangle is height × width, which in this case is radius × half circumference, or πr × r, which is πr².

Finding area using the diameter

The formula for the area of a circle usually uses the radius, but the area can also be found if the diameter is given.

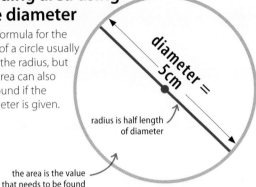

radius is half length of diameter

the area is the value that needs to be found

The formula for the area of a circle is always the same, whatever values are known.

$$\text{area} = \pi\, r^2$$

Substitute the known values into the formula – the radius is 2.5 in this example; half the diameter.

$$\text{area} = 3.14 \times 2.5^2$$

the radius is half the diameter: $5 \div 2 = 2.5$

π is 3.14 to 3 significant figures

Multiply the radius by itself (square it) as shown by the formula – this makes the last multiplication simpler.

$$\text{area} = 3.14 \times 6.25$$

$2.5 \times 2.5 = 6.25$

19.6349... is rounded to 2 decimal places

Make sure the answer is in the right units, cm² in this example, and round it to a suitable number.

$$\text{area} = \mathbf{19.63cm^2}$$

Finding the radius from the area

The formula for area of a circle can also be used to find the radius of a circle if its area is given.

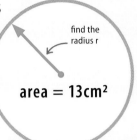

find the radius r

$$\text{area} = 13cm^2$$

The formula for the area of a circle can be used to find the radius if the area is known.

$$\text{area} = \pi\, r^2$$

Substitute the known values into the formula – here the area is 13cm.

$$13 = 3.14 \times r^2$$

divide this side by 3.14

r² was multiplied by 3.14, so divide by 3.14 to isolate r²

Rearrange the formula so r² is on its own on one side – divide both sides by 3.14.

$$\frac{13}{3.14} = r^2$$

Round the answer, and switch the sides so that the unknown, r², is shown first.

r² is shown first

$$r^2 = 4.14$$

4.1380... is rounded to 2 decimal places

Find the square root of the last answer in order to find the value of the radius.

$$r = \sqrt{4.14}$$

2.0342... is rounded to 2 decimal places

Make sure the answer is in the right units (cm² here) and round it to a suitable number.

$$= \mathbf{2.03cm}$$

More complex shapes

When two or more different shapes are put together, the result is called a compound shape. The area of a compound shape can be found by adding the areas of the parts of the shape. In this example, the two different parts are a semicircle, and a rectangle. The total area is 1,414cm² (area of the semicircle, which is ½ × πr², half the area of a circle) + 5,400cm² (the area of the rectangle) = 6,814cm².

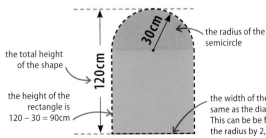

the total height of the shape

120cm

the height of the rectangle is 120 − 30 = 90cm

30cm

the radius of the semicircle

the width of the rectangle is the same as the diameter of the circle. This can be be found by multiplying the radius by 2, 30 × 2 = 60cm

◁ **Compound shapes**
This compound shape consists of a semicircle and a rectangle. Its area can be found using only the two measurements given here.

Angles in a circle

THE ANGLES IN A CIRCLE HAVE A NUMBER OF SPECIAL PROPERTIES.

SEE ALSO

❬ **76–77** Angles
❬ **108–109** Triangles
❬ **130–131** Circles

If angles are drawn to the centre and the circumference from the same two points on the circumference, the angle at the centre is twice the angle at the circumference.

Subtended angles

Any angle within a circle is "subtended" from two points on its circumference – it "stands" on the two points. In both of these examples, the angle at point R is the angle subtended, or standing on, points P and Q. Subtended angles can sit anywhere within the circle.

▷ **Subtended angles**
These circles show how a point is subtended from two other points on the circle's circumference to form an angle. The angle at point R is subtended from points P and Q.

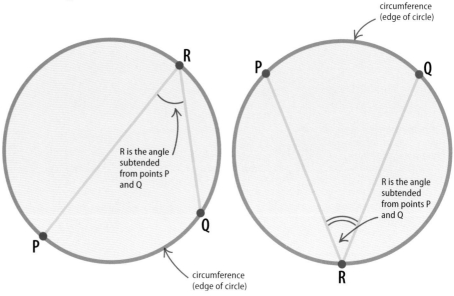

Angles at the centre and at the circumference

When angles are subtended from the same two points to both the centre of the circle and to its circumference, the angle at the centre is always twice the size of the angle formed at the circumference. In this example, both angles R at the circumference and O at the centre are subtended from the same points, P and Q.

$$\frac{\text{angle at}}{\text{centre}} = \frac{2 \times \text{angle at}}{\text{circumference}}$$

▷ **Angle property**
The angles at O and R are both subtended by the points P and Q at the circumference. This means that the angle at O is twice the size of the angle at R.

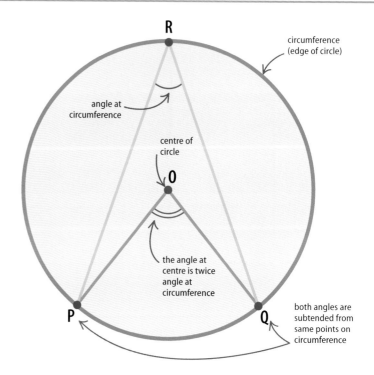

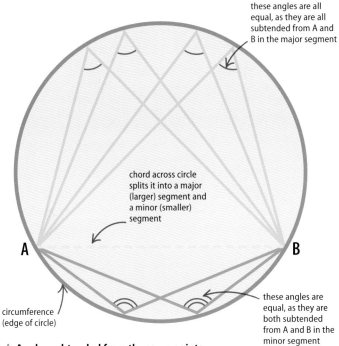

these angles are all equal, as they are all subtended from A and B in the major segment

chord across circle splits it into a major (larger) segment and a minor (smaller) segment

A

B

circumference (edge of circle)

these angles are equal, as they are both subtended from A and B in the minor segment

△ **Angles subtended from the same points**
Angles at the circumference subtended from the same two points in the same segment are equal. Here the angles marked with one red line are equal, as are the angles marked with two red lines.

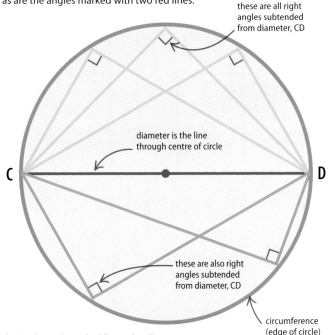

these are all right angles subtended from diameter, CD

diameter is the line through centre of circle

C

D

these are also right angles subtended from diameter, CD

circumference (edge of circle)

△ **Angles subtended from the diameter**
Any angle at the circumference that is subtended from two points either side of the diameter is equal to 90°, which is a right angle.

Proving angle rules in circles

Mathematical rules can be used to prove that the angle at the centre of a circle is twice the size of the angle at the circumference when both the angles are subtended from the same points.

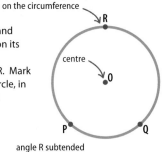

R, P, and Q are 3 points on the circumference

Draw a circle and and mark any 3 points on its circumference, for example, P, Q, and R. Mark the centre of the circle, in this example it is O.

centre

R

O

P

Q

angle R subtended from P and Q

Draw straight lines from R to P, R to Q, O to P, and O to Q. This creates two angles, one at R (the circumference of the circle) and one at O (the centre of the circle). Both are subtended from points P and Q.

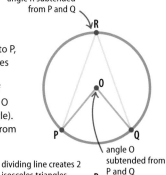

R

O

P

Q

angle O subtended from P and Q

dividing line creates 2 isosceles triangles

Draw a line from R through O, to the other side of the circle. This dividing line creates two isosceles triangles. Isosceles triangles have 2 sides and 2 angles which are the same. In this case, two sides of triangles POR and QOR are formed from 2 radii of the circle.

R

O

P

Q

the angle at O is twice the angle at R

For one triangle the two angles on its base are equal, and labelled A. The exterior angle of this triangle is the sum of the opposite interior angles (A and A), or 2A. Looking at both triangles, it is clear that the angle at O (the centre) is twice the angle at R (the circumference).

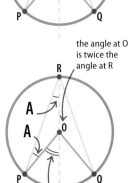

R

A

A

O

P

Q

2A

Chords and cyclic quadrilaterals

A CHORD IS A STRAIGHT LINE JOINING ANY TWO POINTS ON THE CIRCUMFERENCE
OF A CIRCLE. A CYCLIC QUADRILATERAL HAS FOUR CHORDS AS ITS SIDES.

Chords vary in length – the diameter of a circle is also its longest chord. Chords
of the same length are always equal distances from the centre of the circle. The
corners of a cyclic quadrilateral (four-sided shape) touch the circumference of a circle.

SEE ALSO

❮ **122–125** Quadrilaterals

❮ **130–131** Circles

Chords

A chord is a straight line across a circle.
The longest chord of any circle is its
diameter, as the diameter crosses a circle at
its widest point. The perpendicular bisector
of a chord is a line that passes through
its centre at right-angles (90º) to it. The
perpendicular bisector of any chord passes
through the centre of the circle. The distance
of a chord to the centre of a circle is found
by measuring its perpendicular bisector.
If two chords are equal lengths they will
always be the same distance from the
centre of the circle.

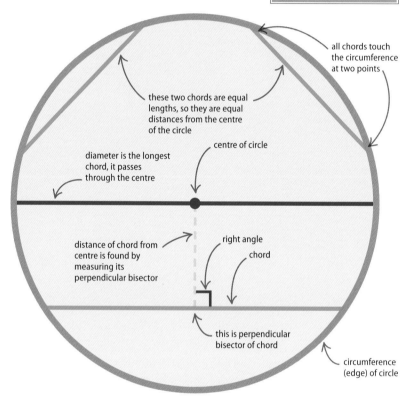

all chords touch
the circumference
at two points

these two chords are equal
lengths, so they are equal
distances from the centre
of the circle

centre of circle

diameter is the longest
chord, it passes
through the centre

distance of chord from
centre is found by
measuring its
perpendicular bisector

right angle

chord

this is perpendicular
bisector of chord

circumference
(edge) of circle

▷ **Chord properties**
This circle shows four chords, the
longest of which is the diameter.
There are two chords that are equal
in length, while the other one is
shown with its perpendicular
bisector (a line that cuts it in half
at right angles).

LOOKING CLOSER

Intersecting chords

When two chords cross, or "intersect", they gain
an interesting property: the two parts of one
chord, either side of where it is split, multiply to
the same value as the answer found by
multiplying the two parts of the other chord.

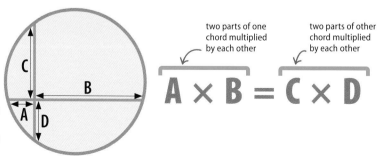

two parts of one
chord multiplied
by each other

two parts of other
chord multiplied
by each other

$$A \times B = C \times D$$

▷ **Crossing chords**
This circle shows two chords, which
cross one another (intersect). One
chord is split into parts A and B, the
other into parts C and D.

Finding the centre of a circle

Chords can be used to find the centre of a circle. To do this, draw any two chords across the circle. Then find the midpoint of each chord, and draw a line through it that is at right angles to the chord (this is a perpendicular bisector). The centre of the circle is where these two lines cross.

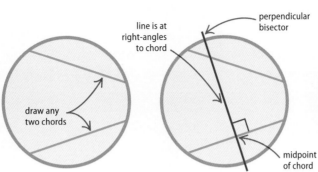
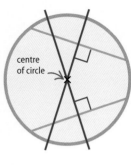

line is at right-angles to chord

perpendicular bisector

centre of circle

draw any two chords

midpoint of chord

First, draw any two chords across the circle of which the centre needs to be found.

Then measure the midpoint of one of the chords, and draw a line through the midpoint at right-angles (90°) to the chord.

Do the same for the other chord. The centre of the circle is the point where the two perpendicular lines cross.

Cyclic quadrilaterals

Cyclic quadrilaterals are four-sided shapes made from chords. Each corner of the shape sits on the circumference of a circle. The interior angles of a cyclic quadrilateral add up to 360°, as they do for all quadrilaterals. The opposite interior angles of a cyclic quadrilateral add up to 180°, and their exterior angles are equal to the opposite interior angles.

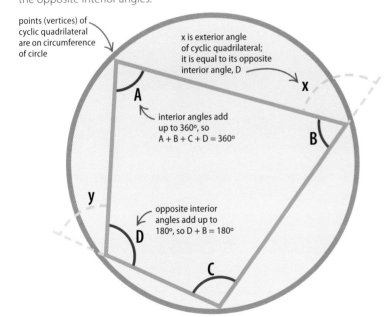

points (vertices) of cyclic quadrilateral are on circumference of circle

x is exterior angle of cyclic quadrilateral; it is equal to its opposite interior angle, D

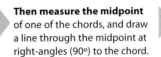

x

A

interior angles add up to 360°, so A + B + C + D = 360°

B

y

opposite interior angles add up to 180°, so D + B = 180°

D

C

$$A + B + C + D = 360°$$

△ **Interior angle sum**
The interior angles of a cyclic quadrilateral always add up to 360°. Therefore, in this example A + B + C + D = 360°.

$$A + C = 180°$$

$$B + D = 180°$$

△ **Opposite angles**
Opposite angles in a cyclic quadrilateral always add up to 180°. In this example, A + C = 180° and B + D = 180°.

exterior angle opposite to angle B

$$y = B$$

exterior angle opposite to angle D

$$x = D$$

△ **Exterior angles**
Exterior angles in cyclic quadrilaterals are equal to the opposite interior angles. Therefore, in this example, y = B and x = D.

△ **Angles in a cyclic quadrilateral**
The four interior angles of this cyclic quadrilateral are A, B, C, and D. Two of the four exterior angles are x and y.

 Tangents

A TANGENT IS A STRAIGHT LINE THAT TOUCHES THE
CIRCUMFERENCE (EDGE) OF A CIRCLE AT A SINGLE POINT.

SEE ALSO

❰ **102–105** Constructions
❰ **120–121** Pythagoras' theorem
❰ **130–131** Circles

What are tangents?

A tangent is a line that extends from a point outside a circle
and touches the edge of the circle in one place, the point of
contact. The line joining the centre of the circle to the point of
contact is a radius, at right-angles (90º) to the tangent. From a
point outside the circle there are two tangents to the circle.

point of contact

circumference

right angle

right angle

tangent

point outside the circle

tangent

point of contact

▷ **Tangent properties**
The lengths of the two tangents from a point
outside a circle to their points of contact are equal.

radius that touches tangent
at point of contact is at
right-angles to tangent

Finding the length of a tangent

A tangent is at right-angles to the radius at the point
of contact, so a right-angled triangle can be created
using the radius, the tangent, and a line between
them, which is the hypotenuse of the triangle.
Pythagoras' theorem can be used to find the length of
any one of the three sides of the right-angled triangle,
if two sides are known.

A

1.5cm

tangent

O

radius

4cm

hypotenuse

P

◁ **Find the tangent**
The tangent, the radius
of the circle, and the
line connecting the
centre of the circle to
point P form a
right-angled triangle.

Pythagoras' theorem shows that the square of the hypotenuse (side
facing the right-angle) of a right-angled triangle is equal to the the
sum of the two squares of the other sides of the triangle.

square of one side   square of other side   square of the hypotenuse

$$a^2 + b^2 = c^2$$

Subsitute the known numbers into the formula. The hypotenuse
is side OP, which is 4cm, and the other known length is the radius,
which is 1.5cm. The side not known is the tangent, AP.

$$1.5^2 + AP^2 = 4^2$$

the value of the
tangent is unknown

Find the squares of the two known sides by multiplying the value
of each by itself. The square of 1.5 is 2.25, and the square of 4 is 16.
Leave the value of the unknown side, AP² as it is.

$1.5 \times 1.5 = 2.25$

$4 \times 4 = 16$

$$2.25 + AP^2 = 16$$

Rearrange the equation to isolate the unknown variable. In this
example the unknown is AP², the tangent. It is isolated by
subtracting 2.25 from both sides of the equation.

subtract 2.25 from
both sides to isolate
the unknown term

2.25 must be subtracted from
both sides to isolate the unknown

$$AP^2 = 16 - 2.25$$

Carry out the subtraction on the right-hand side of the equation.
The value this creates, 13.75, is the squared value of AP, which is the
length of the missing side.

this means AP × AP

$16 - 2.25 = 13.75$

$$AP^2 = 13.75$$

Find the square root of both sides of the equation to find the value
of AP. The square root of AP² is just AP. Use a calculator to find the
square root of 13.75.

the square root
of AP² is just AP

this is the sign for
a square root

$$AP = \sqrt{13.75}$$

Find the square root of the value on the right, and round the answer to a
suitable number of decimal places. This is the length of the missing side.

3.708... is rounded
to 2 decimal places

$$AP = 3.71cm$$

Constructing tangents

To construct a tangent accurately requires a compass and a straight edge. This example shows how to construct two tangents between a circle with centre O and a given point outside the circle, in this case, P.

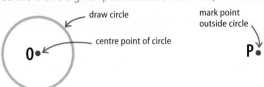

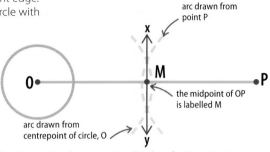

Draw a circle using a compass, and mark the centre O. Also, mark another point outside the circle and label it (in this case P). Construct two tangents to the circle from the point.

Draw a line between O and P, then find its midpoint. Set a compass to just over half OP, and draw two arcs, one from O and one from P. Join the two points where the arcs cross with a straight line (xy). The midpoint is where xy crosses OP.

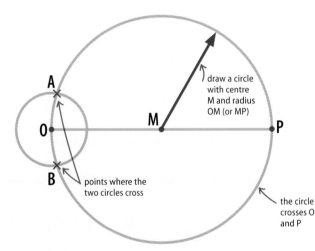

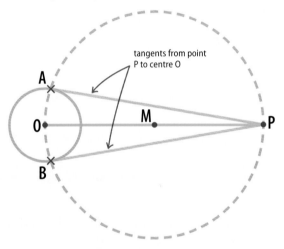

Set the compass to distance OM (or MP which is the same length), and draw a circle with M as its centre. Mark the two points where this new circle intersects (crosses) the circumference of the original circle as A and B.

Finally, join each point where the circles intersect (cross), A and B, with point P. These two lines are the tangents from point P to the circle with centre O. The two tangents are equal lengths.

Tangents and angles

Tangents to circles have some special angle properties. If a tangent touches a circle at B, and a chord, BC, is drawn across the circle from B, an angle is formed between the tangent and the chord at B. If lines (BD and CD) are drawn to the circumference from the ends of the chord, they create an angle at D that is equal to angle B.

▷ **Tangents and chords**
The angle formed between the tangent and the chord is equal to the angle formed at the circumference if two lines are drawn from either end of the chord to meet at a point on the circumference.

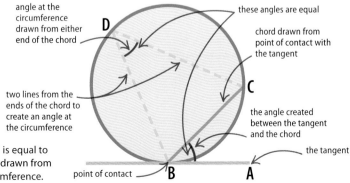

Arcs

AN ARC IS A SECTION OF A CIRCLE'S CIRCUMFERENCE. ITS LENGTH CAN BE FOUND USING ITS RELATED ANGLE AT THE CENTRE OF THE CIRCLE.

SEE ALSO

❮ **48–51** Ratio and proportion
❮ **130–131** Circles
❮ **132–133** Circumference and diameter

What is an arc?

An arc is a part of the circumference of a circle. The length of an arc is in proportion with the size of the angle made at the centre of the circle when lines are drawn from each end of the arc. If the length of an arc is unknown, it can be found using the circumference and this angle. When a circle is split into two arcs, the bigger is called the "major" arc, and the smaller the "minor" arc.

angle created at the centre when two lines are drawn from the ends of the major arc

minor arc

major arc

formula for finding the length of an arc

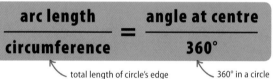

$$\frac{\text{arc length}}{\text{circumference}} = \frac{\text{angle at centre}}{360°}$$

total length of circle's edge   360° in a circle

▷ **Arcs and angles**
This diagram shows two arcs: one major, one minor, and their angles at the centre of the circle.

angle created at the centre when two lines are drawn from the ends of the minor arc

Finding the length of an arc

The length of an arc is a proportion of the whole circumference of the circle. The exact proportion is the ratio between the angle formed from each end of the arc at the centre of the circle, and 360°, which is the total number of degrees around the central point. This ratio is part of the formula for the length of an arc.

120°

◁ **Find the arc length**
This circle has a circumference of 10cm. Find the length of the arc that forms an angle of 120° at the centre of the circle.

circumference is 10cm

Take the formula for finding the length of an arc. The formula uses the ratios between arc length and circumference, and between the angle at the centre of the circle and 360° (total number of degrees).

$$\frac{\text{arc length}}{\text{circumference}} = \frac{\text{angle at centre}}{360°}$$

Substitute the numbers that are known into the formula. In this example, the circumference is known to be 10cm, and the angle at the centre of the circle is 120°; 360° stays as it is.

$$\frac{\text{arc length}}{10} = \frac{120}{360}$$

this side has been multiplied by 10 to leave arc length on its own (÷10 × 10 cancels out)

this side has also been multiplied by 10, as what is done to one side must be done to the other

Rearrange the equation to isolate the unknown value – the arc length – on one side of the equals sign. In this example the arc length is isolated by multiplying both sides by 10.

$$\text{arc length} = \frac{10 \times 120}{360}$$

3.333... is rounded to 2 decimal places

Multiply 10 by 120 and divide the answer by 360 to get the value of the arc length. Then round the answer to a suitable number of decimal places.

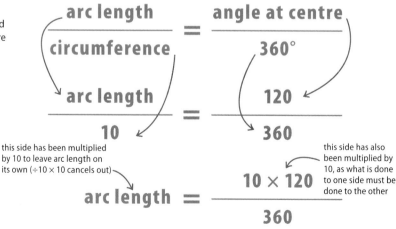

$$C = 3.33\text{cm}$$

Sectors

A SECTOR IS A SLICE OF A CIRCLE'S AREA. ITS AREA CAN BE FOUND
USING THE ANGLE IT CREATES AT THE CENTRE OF THE CIRCLE.

SEE ALSO

❮ **48–51** Ratio and proportion

❮ **130–131** Circles

❮ **132–133** Circumference and diameter

What is a sector?

A sector of a circle is the space between two radii and one arc. The area of a sector depends on the size of the angle between the two radii at the centre of the circle. If the area of a sector is unknown, it can be found using this angle and the area of the circle. When a circle is split into two sectors, the bigger is called the "major" sector, and the smaller the "minor" sector.

$$\frac{\text{area of sector}}{\text{area of circle}} = \frac{\text{angle at centre}}{360°}$$

↖ formula for finding the area of a sector

minor arc

angle at the centre created by the two radii of the minor sector

minor sector

angle at the centre created by the two radii of the major sector

major sector

major arc

▷ **Sectors and angles**
This diagram shows two sectors: one major, one minor, and their angles at the centre of the circle.

Finding the area of a sector

The area of a sector is a proportion, or part, of the area of the whole circle. The exact proportion is the ratio of the angle formed between the two radii that are the edges of the sector and 360°. This ratio is part of the formula for the area of a sector.

45°

angle formed by sector

◁ **Find the sector area**
This circle has an area of 7cm². Find the area of the sector that forms a 45° angle at the centre of the circle.

the area of the circle is 7cm²

Take the formula for finding the area of a sector. The formula uses the ratios between the area of a sector and the area of the circle, and between the angle at the centre of the circle and 360°.

$$\frac{\text{area of sector}}{\text{area of circle}} = \frac{\text{angle at centre}}{360°}$$

total number of degrees in circle

Substitute the numbers that are known into the formula. In this example, the area is known to be 7cm², and the angle at the centre of the circle is 45°. The total number of degrees in a circle is 360°.

$$\frac{\text{area of sector}}{7} = \frac{45}{360}$$

Rearrange the equation to isolate the unknown value – the area of the sector – on one side of the equals sign. In this example, this is done by multiplying both sides by 7.

this side has been multiplied by 7 to leave the area of a sector on its own (÷7 ×7 cancels out)

this side has also been multiplied by 7

$$\text{area of sector} = \frac{45 \times 7}{360}$$

Multiply 45 by 7 and divide the answer by 360 to get the area of the sector. Round the answer to a suitable number of decimal places.

0.875 is rounded to 2 decimal places

$$C = 0.88\text{cm}^2$$

Solids

A SOLID IS A THREE-DIMENSIONAL SHAPE.

SEE ALSO

‹ 126–129 Polygons	
Volume	146–147 ›
Surface area	148–149 ›

Solids are objects with three dimensions: width, length, and height. They also have surface areas and volumes.

Prisms

Many common solids are polyhedrons – three-dimensional shapes with flat surfaces and straight edges. Prisms are a type of polyhedron made up of two parallel shapes of exactly the same shape and size, which are connected by faces. In the example to the right, the parallel shapes are pentagons, joined by rectangular faces. Usually a prism is named after the shape of its ends, so a cylinder is a circular prism. A cuboid is a prism whose parallel shapes are rectangles, so it is known as a rectangular prism.

▷ **A prism**
The cross-section of this prism is a pentagon (a shape with five sides), so it is called a pentagonal prism.

◁ **Volume**
The amount of space that a solid occupies is called its volume.

edge
a line where surfaces meet

height
distance from top to bottom

width
horizontal distance at right angles to length

cross-section of pentagonal prism is a pentagon

a pentagon is a shape with five sides

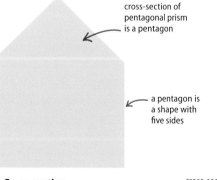

△ **Cross-section**
A cross-section is the shape made when an object is sliced from top to bottom.

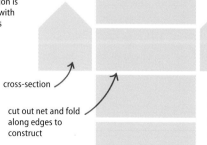

cross-section

cut out net and fold along edges to construct

this net forms a shape with seven faces

◁ **Surface area**
The surface area of a solid is the total area of its net – a two-dimensional shape, or plan, that forms the solid if it is folded up.

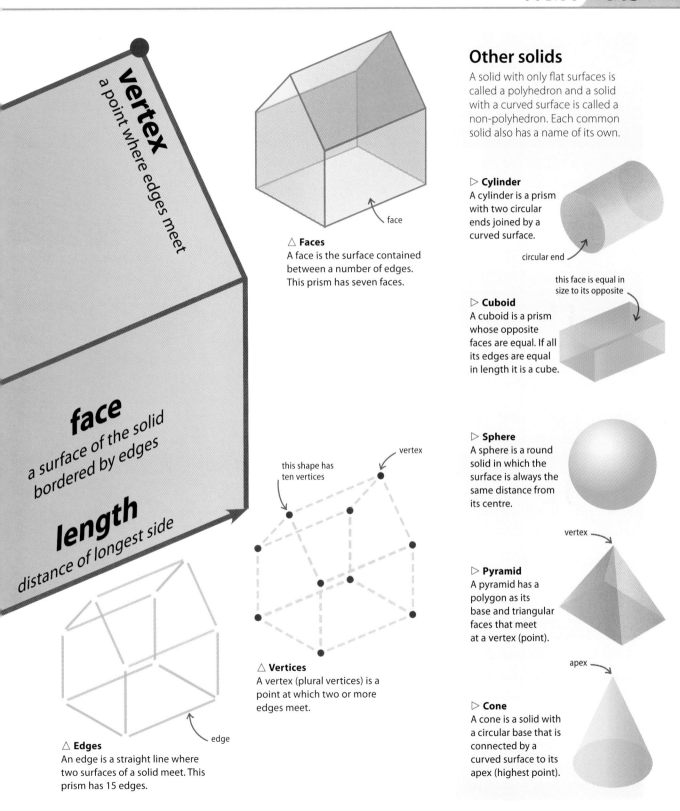

vertex
a point where edges meet

face
a surface of the solid
bordered by edges

length
distance of longest side

△ **Faces**
A face is the surface contained
between a number of edges.
This prism has seven faces.

face

△ **Edges**
An edge is a straight line where
two surfaces of a solid meet. This
prism has 15 edges.

edge

△ **Vertices**
A vertex (plural vertices) is a
point at which two or more
edges meet.

this shape has
ten vertices

vertex

Other solids
A solid with only flat surfaces is
called a polyhedron and a solid
with a curved surface is called a
non-polyhedron. Each common
solid also has a name of its own.

▷ **Cylinder**
A cylinder is a prism
with two circular
ends joined by a
curved surface.

circular end

▷ **Cuboid**
A cuboid is a prism
whose opposite
faces are equal. If all
its edges are equal
in length it is a cube.

this face is equal in
size to its opposite

▷ **Sphere**
A sphere is a round
solid in which the
surface is always the
same distance from
its centre.

▷ **Pyramid**
A pyramid has a
polygon as its
base and triangular
faces that meet
at a vertex (point).

vertex

▷ **Cone**
A cone is a solid with
a circular base that is
connected by a
curved surface to its
apex (highest point).

apex

 # Volumes

THE AMOUNT OF SPACE WITHIN A THREE-DIMENSIONAL SHAPE.

SEE ALSO

❰ **28–29** Units of measurement

❰ **144–145** Solids

❰ **148–149** Surface area

Solid space

When measuring volume, unit cubes, also called cubic units are used, for example, cm³ and m³. An exact number of unit cubes fits neatly into some types of three-dimensional shapes, also known as solids, such as a cube, but for most solids, for example, a cylinder, this is not the case. Formulas are used to find the volumes of solids. Finding the area of the base, or the cross-section, of a solid is the key to finding its volume. Each solid has a different cross-section.

▷ **Unit cubes**
A unit cube has sides that are of equal size. A 1cm cube has a volume of $1 \times 1 \times 1$ cm, or 1cm³. The space within a solid can be measured by the number of unit cubes that can fit inside. This cuboid has a volume of $3 \times 2 \times 2$ cm, or 12cm³.

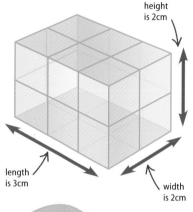

height is 2cm

length is 3cm

width is 2cm

Finding the volume of a cylinder

A cylinder is made up from a rectangle and two circles. Its volume is found by multiplying the area of a circle with the length, or height, of the cylinder.

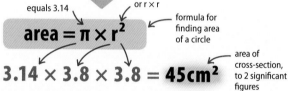

$$\text{volume} = \pi \times r^2 \times l$$

formula for finding volume of a cylinder

The formula for the volume of a cylinder uses the formula for the area of a circle multiplied by the length of the cylinder.

equals 3.14

or $r \times r$

$$\text{area} = \pi \times r^2$$

formula for finding area of a circle

$$3.14 \times 3.8 \times 3.8 = 45\text{cm}^2$$

area of cross-section, to 2 significant figures

First, find the area of the cylinder's cross-section using the formula for finding the area of a circle. Insert the values given on the illustration of the cylinder below.

$$\text{volume} = \text{area} \times \text{length}$$

$$45 \times 12 = 544\text{cm}^3$$

Next, multiply the area by the length of the cylinder to find its volume.

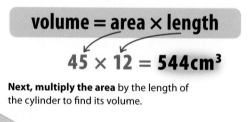

LENGTH =12cm

RADIUS = 3.8cm

area of a cross-section

▷ **Circular cross-section**
The base of a cylinder is a circle. When a cylinder is sliced widthways, the circles created are identical and so a cylinder is said to have a circular cross-section.

Finding the volume of a cuboid

A cuboid has six flat sides and all of its faces are rectangles. Multiply the length by the width by the height to find the volume of a cuboid.

formula also written
v = h × w × l, or v = hwl

volume = length × width × height

$$4.3 \times 2.2 \times 1.7 = \textbf{16cm}^3$$

answer rounded to 2 significant figures

▷ **Multiply lengths of the sides**
This cuboid has a length of 4.3cm, a width of 2.2cm, and a height of 1.7cm. Multiply these measurements to find its volume.

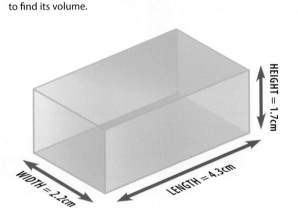

HEIGHT = 1.7cm

WIDTH = 2.2cm

LENGTH = 4.3cm

Finding the volume of a cone

Multiply the distance from the tip of the cone to the centre of its base (the vertical height) with the area of its base (the area of a circle), and then multiply by ⅓.

also called the perpendicular height

volume $= \dfrac{1}{3} \times \pi \times r^2 \times$ **vertical height**

$$\dfrac{1}{3} \times 3.14 \times 2 \times 2 \times 4.3 = \textbf{18cm}^3$$

▷ **Using the formula**
To find the volume of this cone, multiply together ⅓, π, the radius squared, and the vertical height.

answer rounded to 2 significant figures

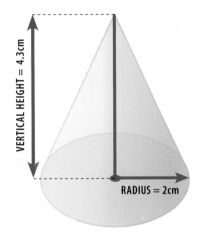

VERTICAL HEIGHT = 4.3cm

RADIUS = 2cm

Finding the volume of a sphere

The radius is the only measurement needed to find the volume of a sphere. This sphere has a radius of 2.5cm.

multiply radius by itself twice

volume $= \dfrac{4}{3} \times \pi \times r^3$

$$\dfrac{4}{3} \times 3.14 \times 2.5 \times 2.5 \times 2.5 = \textbf{65cm}^3$$

▷ **Using the formula**
To find the volume of this sphere, multiply together ⁴⁄₃, π, and the radius cubed (the radius multiplied by itself twice).

answer rounded to 2 significant figures

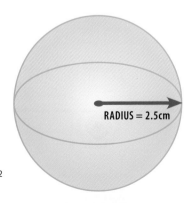

RADIUS = 2.5cm

Surface area of solids

SURFACE AREA IS THE SPACE OCCUPIED BY A SHAPE'S OUTER SURFACES.

SEE ALSO

❮ **28–29** Units of measurement

❮ **144–145** Solids

❮ **146–147** Volume

For most solids, surface area can be found by adding together the areas of its faces. The sphere is the exception, but there is an easy formula to use.

Surfaces of shapes

For all solids with straight edges, surface area can be found by adding together the areas of all the solid's faces. One way to do this is to imagine taking apart and flattening out the solid into two-dimensional shapes. It is then straightforward to work out and add together the areas of these shapes. A diagram of a flattened and opened out shape is known as its net.

▷ **Cylinder**
A cylinder has two flat faces and a curved surface. To create its net, the flat surfaces are separated and the curved surface opened up.

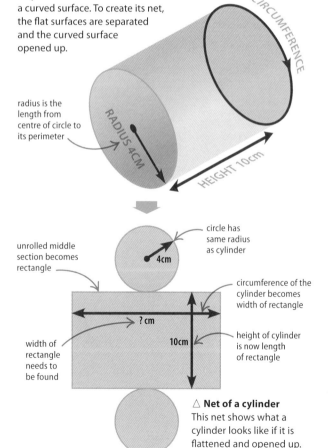

radius is the length from centre of circle to its perimeter

RADIUS 4CM

CIRCUMFERENCE

HEIGHT 10cm

unrolled middle section becomes rectangle

circle has same radius as cylinder

4cm

circumference of the cylinder becomes width of rectangle

? cm

10cm

width of rectangle needs to be found

height of cylinder is now length of rectangle

△ **Net of a cylinder**
This net shows what a cylinder looks like if it is flattened and opened up. It consists of a rectangle and two circles.

Finding the surface area of a cylinder

Breaking the cylinder down into its component parts creates a rectangle and two circles. To find the total surface area, work out the area of each of these and add them together.

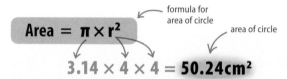

formula for area of circle

area of circle

$$\text{Area} = \pi \times r^2$$

$$3.14 \times 4 \times 4 = 50.24\,\text{cm}^2$$

The area of the circles can be worked out using the known radius and the formula for the area of a circle. π (pi) is usually shortened to 3.14, and area is always expressed in square units.

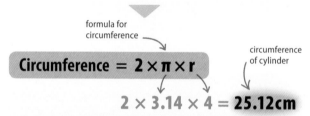

formula for circumference

circumference of cylinder

$$\text{Circumference} = 2 \times \pi \times r$$

$$2 \times 3.14 \times 4 = 25.12\,\text{cm}$$

Before the area of the rectangle can be found, it is necessary to work out its width – the circumference of the cylinder. This is done using the known radius and the formula for circumference.

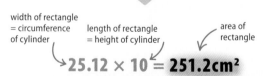

width of rectangle = circumference of cylinder

length of rectangle = height of cylinder

area of rectangle

$$25.12 \times 10 = 251.2\,\text{cm}^2$$

The area of the rectangle can now be found by using the formula for the area of rectangle (length × width).

surface area of cylinder

$$50.24 + 50.24 + 251.2 = 351.68\,\text{cm}^2$$

The surface area of the cylinder is found by adding together the areas of the three shapes that make up its net – two circles and a rectangle.

CFE AREA OF SOLIDS
CFE AREA OF SOLIDS

Finding the surface area of a cuboid

A cuboid is made up of three different pairs of rectangles, here labelled A, B, and C. The surface area of a cuboid is the sum of the areas of all its faces.

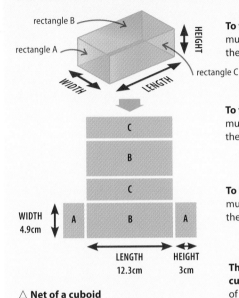

rectangle B

rectangle A

HEIGHT

WIDTH

LENGTH

rectangle C

C

B

C

WIDTH
4.9cm

A | B | A

LENGTH
12.3cm

HEIGHT
3cm

△ **Net of a cuboid**
The net of a cuboid is made up of three different pairs of rectangles.

To find the area of rectangle A, multiply together its two sides: the cuboid's height and width.

▽

To find the area of rectangle B, multiply together its two sides: the cuboid's length and width.

▽

To find the area of rectangle C, multiply together its two sides: the cuboid's height and length.

▽

The surface area of the cuboid is the total of the areas of its sides – twice area A, added to twice area B, added to twice area C.

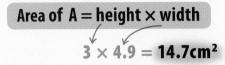

Area of A = height × width

$3 \times 4.9 = \textbf{14.7cm}^2$

Area of B = length × width

$12.3 \times 4.9 = \textbf{60.27cm}^2$

Area of C = height × length

$3 \times 12.3 = \textbf{36.9cm}^2$

brackets used to separate operations

$(2 \times A) + (2 \times B) + (2 \times C)$

$(2 \times 14.7) + (2 \times 60.27) + (2 \times 36.9)$

$= \textbf{223.74cm}^2$

Finding the surface area of a cone

A cone is made up of two parts – a circular base and a cone shape. Formulas are used to find the areas of the two parts, which are then added together to give the surface area.

Area = π × r × h

slant height

surface area of cone without base

$3.14 \times 3.9 \times 9 = \textbf{110.21cm}^2$

π × r²

formula for area of a circle

surface area of base

$3.14 \times 3.9 \times 3.9 = \textbf{47.76cm}^2$

total surface area of cone

$110.21 + 47.76 = \textbf{157.97cm}^2$

▷ **Cone**
Find the surface area of a cone by using formulas to find the area of the cone shape and the area of the base, and adding the two.

To find the area of the cone, multiply π by the radius and slant length.

▽

To find the area of the base, use the formula for the area of a circle, $\pi \times r^2$.

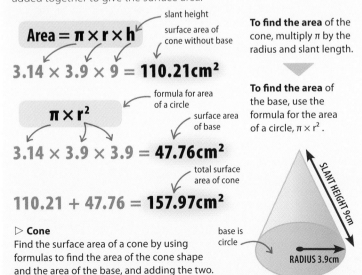

SLANT HEIGHT 9cm

base is circle

RADIUS 3.9cm

Finding the surface area of a sphere

Unlike many other solid shapes, a sphere cannot be unrolled or unfolded. Instead, a formula is used to find its surface area.

Area = 4 × π × r²

formula for the surface area of sphere

$4 \times 3.14 \times 17 \times 17$

$= \textbf{3,629.84cm}^2$

▷ **Sphere**
The formula for the surface area of a sphere is the same as 4 times the formula for the area of a circle (πr^2). This means that the surface area of a sphere is equal to the surface area of 4 circles with the same radius.

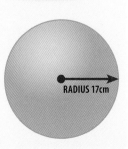

RADIUS 17cm

Trigonometry

What is trigonometry?

TRIGONOMETRY DEALS WITH THE RELATIONSHIPS BETWEEN THE
SIZES OF ANGLES AND LENGTHS OF SIDES IN TRIANGLES.

SEE ALSO

❮ **48–51** Ratio and proportion
❮ **117–119** Similar triangles

Corresponding triangles

Trigonometry uses comparisons of the lengths of the sides of
similar right-angled triangles (which have the same shape but
different sizes) to find the sizes of unknown angles and sides.
This diagram shows the Sun creating shadows of a person and
a building, which form two similar triangles. By measuring the
shadows, the height of the person, which is known, can be
used to find the height of the building, which is unknown.

▽ **Similar triangles**
The shadows the sun makes
of the person and the
building create two
corresponding triangles.

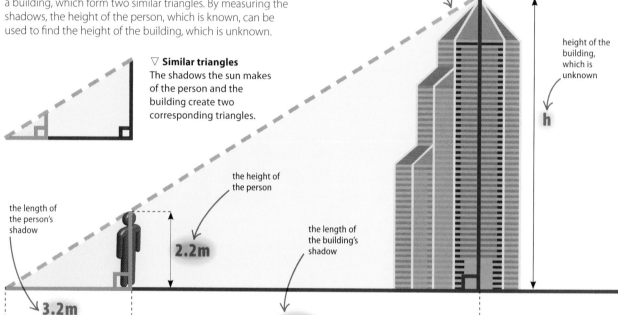

the Sun's rays
create shadows of
the person and the
building

the sun

height of the
building,
which is
unknown

h

the height of
the person

the length of
the building's
shadow

the length of
the person's
shadow

2.2m

3.2m

58m

▷ **The ratio between** corresponding sides of similar triangles
is equal, so the building's height divided by the person's
height equals the length of the building's shadow divided by
the length of the person's shadow.

$$\frac{\text{height of building}}{\text{height of person}} = \frac{\text{length of building's shadow}}{\text{length of person's shadow}}$$

▷ **Substitute the values** from the diagram into this
equation. This leaves only one unknown – the height of
the building (h) – which is found by rearranging the
equation.

the value of h is
unknown

$$\frac{h}{2.2} = \frac{58}{3.2}$$

whatever is done to one
side of the equation must
be done to the other, so
this side must also be
multiplied by 2.2

▷ **Rearrange the equation** to leave h (the height of the
building) on its own. This is done by multiplying both
sides of the equation by 2.2, then cancelling out the two
2.2s on the left side, leaving just h.

this side has been
multiplied by 2.2 to
cancel out the ÷ 2.2
and isolate h

$$h = \frac{58}{3.2} \times 2.2$$

the answer is
rounded to 2
decimal places

▷ **Work out the right side** of the equation to find the
value of h, which is the height of the building.

$$h = 39.88m$$

 # Using formulas in trigonometry

SEE ALSO

❮ **48–51** Ratio and proportion

❮ **117–119** Similar triangles

Finding missing sides **154–155** ❯

Finding missing angles **156–157** ❯

TRIGONOMETRY FORMULAS CAN BE USED TO WORK OUT THE LENGTHS OF SIDES AND SIZES OF ANGLES IN TRIANGLES.

Right-angled triangles

The sides of these triangles are called the hypotenuse, opposite, and adjacent. The hypotenuse is always the side opposite the right angle. The names of the other two sides depend on where they are in relation to the particular angle specified.

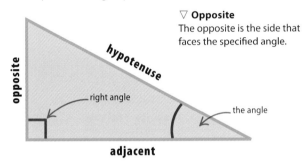

▽ **Opposite**
The opposite is the side that faces the specified angle.

▽ **Adjacent**
The adjacent is the shorter side next to the specified angle.

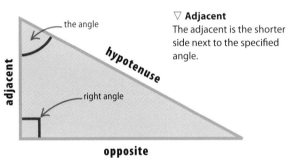

Trigonometry formulas

There are three basic formulas used in trigonometry. "A" stands in for the angle that is being found (this may also sometimes be written as θ). The formula to use depends on the sides of the triangle that are known.

$$\sin A = \frac{\text{opposite}}{\text{hypotenuse}}$$

△ **The sine formula**
The sine formula is used when the lengths of the opposite and hypotenuse are known.

$$\cos A = \frac{\text{adjacent}}{\text{hypotenuse}}$$

△ **The cosine formula**
The cosine formula is used when the lengths of the adjacent and hypotenuse are known.

$$\tan A = \frac{\text{opposite}}{\text{adjacent}}$$

△ **The tangent formula**
The tangent formula is used when the lengths of the opposite and adjacent are known.

Using a calculator

The values of sine, cosine, and tangent are set for each angle. Calculators have buttons that retrieve these values. Use them to find the sine, cosine, or tangent of a particular angle.

△ **Sine, cosine, and tangent**
Press the sine, cosine, or tangent button then enter the value of the angle to find its sine, cosine, or tangent.

 then

△ **Inverse sine, cosine, and tangent**
Press the shift button then the sin, cosine, or tangent button, then enter the value of the sine, cosine, or tangent to find the inverse (the angle in degrees).

Finding missing sides

GIVEN AN ANGLE AND THE LENGTH OF ONE SIDE OF A RIGHT-ANGLED TRIANGLE, THE OTHER SIDES CAN BE FOUND.

SEE ALSO

❮ **152–153** What is trigonometry?

Finding missing angles **156–157** ❯

Formulas **169–171** ❯

The trigonometry formulas can be used to find a length in a right-angled triangle if one angle (other than the right-angle) and one other side are known. Use a calculator to find the sine, cosine, or tangent of an angle.

▽ **Calculator buttons**
These calculator buttons recall the value of sine, cosine, and tangent for any value entered.

Which formula?

The formula to use depends on what information is known. Choose the formula that contains the known side as well as the side that needs to be found. For example, use the sine formula if the length of the hypotenuse is known, one angle other than the right angle is known, and the length of the side opposite the given angle needs to be found.

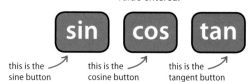

this is the sine button

this is the cosine button

this is the tangent button

$$\sin A = \frac{\text{opposite}}{\text{hypotenuse}}$$

△ **The sine formula**
This formula is used if one angle, and either the side opposite it or the hypotenuse are given.

$$\cos A = \frac{\text{adjacent}}{\text{hypotenuse}}$$

△ **The cosine formula**
Use this formula if one angle and either the side adjacent to it or the hypotenuse are known.

$$\tan A = \frac{\text{opposite}}{\text{adjacent}}$$

△ **The tangent formula**
This formula is used if one angle and either the side opposite it or adjacent to it are given.

Using the sine formula

In this right-angled triangle, one angle other than the right-angle is known, as is the length of the hypotenuse. The length of the side opposite the angle is missing and needs to be found.

Choose the right formula – as the hypotenuse is known and the value for the opposite side is what needs to be found, use the sine formula.

▽

Substitute the known values into the sine formula.

▽

Rearrange the formula to make the unknown (x) the subject by multiplying both sides by 7.

▽

Use a calculator to find the value of sin 37° – press the sin button then enter 37.

▽

Round the answer to a suitable size.

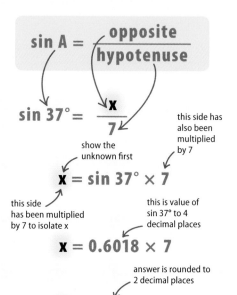

$$\sin A = \frac{\text{opposite}}{\text{hypotenuse}}$$

$$\sin 37° = \frac{x}{7}$$

show the unknown first

this side has also been multiplied by 7

$$x = \sin 37° \times 7$$

this side has been multiplied by 7 to isolate x

this is value of sin 37° to 4 decimal places

$$x = 0.6018 \times 7$$

answer is rounded to 2 decimal places

$$x = 4.21\text{cm}$$

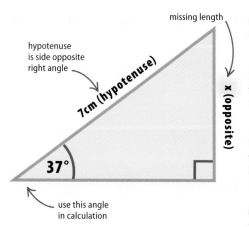

missing length

hypotenuse is side opposite right angle

7cm (hypotenuse)

x (opposite)

37°

use this angle in calculation

Using the cosine formula

In this right-angled triangle, one angle other than the right-angle is known, as is the length of the side adjacent to it. The hypotenuse is the missing side that needs to be found.

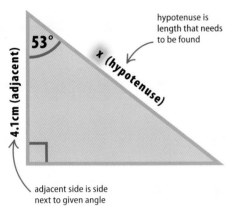

53°

4.1cm (adjacent)

x (hypotenuse)

hypotenuse is length that needs to be found

adjacent side is side next to given angle

Choose the right formula – as the side adjacent to the angle is known and the value of the hypotenuse is missing, use the cosine formula.

$$\cos A = \frac{\text{adjacent}}{\text{hypotenuse}}$$

Substitute the known values into the formula.

$$\cos 53° = \frac{4.1}{x}$$

Rearrange to make x the subject of the equation – first multiply both sides by x.

this side has been multiplied by x

this side has also been multiplied by x, leaving 4.1 on its own

$$\cos 53° \times x = 4.1$$

Divide both sides by cos 53° to make x the subject of the equation.

ths side has been divided by cos 53° to isolate x

this side has also been divided by cos 53°

$$x = \frac{4.1}{\cos 53°}$$

Use a calculator to find the value of cos 53° – press the cos button then enter 53.

value of cos 53° is rounded to 4 decimal places

$$x = \frac{4.1}{0.6018}$$

Round the answer to a suitable size.

answer is rounded to 2 decimal places

$$x = 6.81\text{cm}$$

Using the tangent formula

In this right-angled triangle, one angle other than the right-angle is known, as is the length of the side adjacent to it. Find the length of the side opposite the angle.

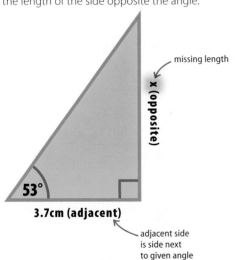

53°

3.7cm (adjacent)

x (opposite)

missing length

adjacent side is side next to given angle

Choose the right formula – as the side adjacent to the angle given are known and the opposite side is sought, use the tangent formula.

$$\tan A = \frac{\text{opposite}}{\text{adjacent}}$$

Substitute the known values into the tangent formula.

$$\tan 53° = \frac{x}{3.7}$$

this side has also been multiplied by 3.7

Rearrange to make x the subject by multiplying both sides by 3.7.

show the unknown first

this side has been multiplied by 3.7 to isolate x

$$x = \tan 53° \times 3.7$$

Use a calculator to find the value of tan 53° – press the tan button then enter 53.

value of tan 53° is rounded to 4 decimal places

$$x = 1.3270 \times 3.7$$

Round the answer to a suitable size.

the answer is rounded to 2 decimal places

$$x = 4.91\text{cm}$$

Finding missing angles

IF THE LENGTHS OF TWO SIDES OF A RIGHT-ANGLED TRIANGLE ARE
KNOWN, ITS MISSING ANGLES CAN BE FOUND.

SEE ALSO
❮ 64–65 Using a calculator
❮ 152–153 What is trigonometry?
❮ 154–155 Finding missing sides
Formulas 169–171 ❯

To find the missing angles in a right-angled triangle, the inverse sine,
cosine, and tangent are used. Use a calculator to find these values.

Which formula?

Choose the formula that contains the pair of sides that are
given in an example. For instance, use the sine formula if the
lengths of the hypotenuse and the side opposite the unknown
angle are known, and the cosine formula if the lengths of the
hypotenuse and the side next to the angle are given.

▽ **Calculator functions**
To find the inverse values of sine, cosine,
and tangent, press shift before sine,
cosine, or tangent.

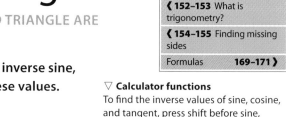

sin^{-1} cos^{-1} tan^{-1}

SHIFT then **sin** **cos** **tan**

this is the sine button   this is the cosine button   this is the tangent button

$$\sin A = \frac{opposite}{hypotenuse}$$

△ **The sine formula**
Use the sine formula if the lengths of the
hypotenuse and the side opposite the
missing angle are known.

$$\cos A = \frac{adjacent}{hypotenuse}$$

△ **The cosine formula**
Use the cosine formula if the lengths of the
hypotenuse and the side adjacent (next to)
to the missing angle are known.

$$\tan A = \frac{opposite}{adjacent}$$

△ **The tangent formula**
Use the tangent formula if the lengths of
the sides opposite and adjacent to the
missing angle are known.

Using the sine formula

In this right-angled triangle the hypotenuse and the side opposite
angle A are known. Use the sine formula to find the size of angle A.

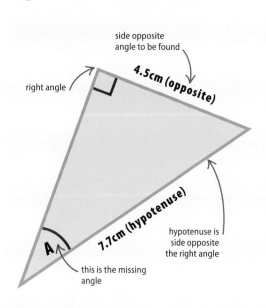

side opposite
angle to be found

4.5cm (opposite)

right angle

7.7cm (hypotenuse)

hypotenuse is
side opposite
the right angle

A

this is the missing
angle

Choose the right formula – in
this example the hypotenuse
and the side opposite the
missing angle, A, are known,
so use the sine formula.

▼

Substitute the known values
into the sine formula.

▼

Work out the value of sin A
by dividing the opposite side
by the hypotenuse.

▼

Find the value of the angle by
using the inverse sine function
on a calculator.

▼

Round the answer to a
suitable size. This is the value
of the missing angle.

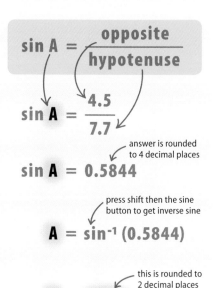

$$\sin A = \frac{opposite}{hypotenuse}$$

$$\sin A = \frac{4.5}{7.7}$$

answer is rounded
to 4 decimal places

$$\sin A = 0.5844$$

press shift then the sine
button to get inverse sine

$$A = \sin^{-1}(0.5844)$$

this is rounded to
2 decimal places

$$A = 35.76°$$

Using the cosine formula

In this right-angled triangle the hypotenuse and the side adjacent to angle A are known. Use the cosine formula to find the size of angle A.

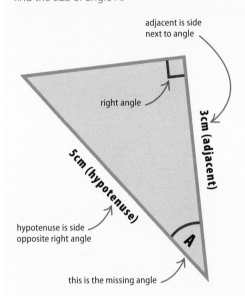

adjacent is side
next to angle

right angle

3 cm (adjacent)

5 cm (hypotenuse)

hypotenuse is side
opposite right angle

A

this is the missing angle

Choose the right formula – in this example the hypotenuse and the side adjacent to the mssing angle, A, are known, so use the cosine formula.

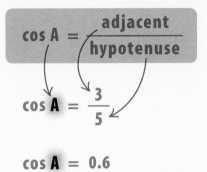

$$\cos A = \frac{adjacent}{hypotenuse}$$

Substitute the known values into the formula.

$$\cos A = \frac{3}{5}$$

Work out the value of cos A by dividing the adjacent side by the length of the hypotenuse.

$$\cos A = 0.6$$

Find the value of the angle by using the inverse cosine function on a calculator.

press shift then cosine button to get inverse cosine

$$A = \cos^{-1}(0.6)$$

Round the answer to a suitable size. This is the value of the missing angle.

answer is rounded to 2 decimal places

$$A = 53.13°$$

Using the tangent formula

In this right-angled triangle the sides opposite and adjacent to angle A are known. Use the tangent formula to find the size of angle A.

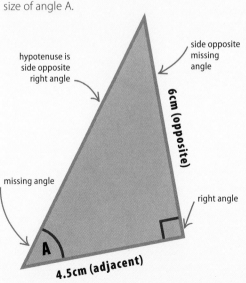

hypotenuse is
side opposite
right angle

side opposite
missing
angle

6 cm (opposite)

missing angle

right angle

A

4.5 cm (adjacent)

Choose the right formula – here the sides opposite and adjacent to the missing angle, A, are known, so use the tangent formula.

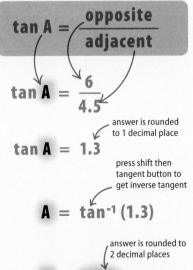

$$\tan A = \frac{opposite}{adjacent}$$

Substitute the known values into the tangent formula.

$$\tan A = \frac{6}{4.5}$$

answer is rounded to 1 decimal place

Work out the value of tan A by dividing the opposite by the adjacent.

$$\tan A = 1.3$$

press shift then tangent button to get inverse tangent

Find the value of the angle by using the inverse tangent function on a calculator.

$$A = \tan^{-1}(1.3)$$

answer is rounded to 2 decimal places

Round the answer to a suitable size. This is the value of the missing angle.

$$A = 52.43°$$

Algebra

 What is algebra?

ALGEBRA IS A BRANCH OF MATHEMATICS IN WHICH LETTERS AND SYMBOLS ARE USED TO REPRESENT NUMBERS AND THE RELATIONSHIPS BETWEEN NUMBERS.

Algebra is widely used in maths, in sciences such as physics, as well as in other areas, such as economics. Formulas for solving a wide range of problems are often given in algebraic form.

Using letters and symbols

Algebra uses letters and symbols. Letters usually represent numbers, and symbols represent operations, such as addition and subtraction. This allows relationships between quantities to be written in a short, generalized way, eliminating the need to give individual specific examples containing actual values. For instance, the volume of a cuboid can be written as lwh (which means length × width × height), enabling the volume of any cuboid to be found once its dimensions are known.

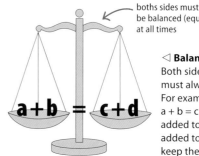

boths sides must be balanced (equal) at all times

◁ **Balancing**
Both sides of an equation must always be balanced. For example, in the equation a + b = c + d, if a number is added to one side, it must be added to the other side to keep the equation balanced.

TERM
The parts of an algebraic expression that are separated by symbols for operations, such as + and −. A term can be a number, a letter, or a combination of both

OPERATION
A procedure carried out on the terms of an algebraic expression, such as addition, subtraction, multiplication, and division

VARIABLE
An unknown number or quantity represented by a letter

EXPRESSION
An expression is a statement written in algebraic form, 2 + b in the example above. An expression can contain any combination of numbers, letters, and symbols (such as + for addition)

△ **Algebraic equation**
An equation is a mathematical statement that two things are equal. In this example, the left side (2 + b) is equal to the right side (8).

REAL WORLD

Algebra in everyday life

Although algebra may seem abstract, with equations consisting of strings of symbols and letters, it has many applications in everyday life. For example, an equation can be used to find out the area of something, such as a tennis court.

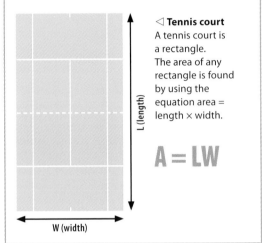

◁ **Tennis court**
A tennis court is a rectangle. The area of any rectangle is found by using the equation area = length × width.

L (length)

W (width)

$$A = LW$$

EQUALS
The equals sign means that the two sides of the equation balance each other

CONSTANT
A number with a value that is always the same

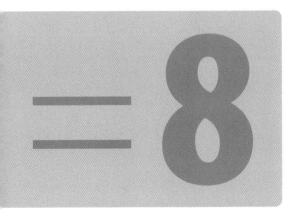

THE ANSWER IS:
b = 6

BASIC RULES OF ALGEBRA

Like other areas of maths, algebra has rules that must be followed to get the correct answer. For example, one rule is about the order in which operations must be done.

Addition and subtraction

Terms can be added together in any order in algebra. However, when subtracting, the order of the terms must be kept as it was given.

$$a + b = b + a$$

△ **Two terms**
When adding together two terms, it is possible to start with either term.

$$(a + b) + c = a + (b + c)$$

△ **Three terms**
As with adding two terms, three terms can be added together in any order.

Multiplication and division

Multiplying terms in algebra can be done in any order, but when dividing the terms must be kept in the order they were given.

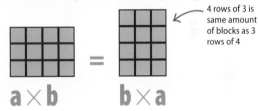

4 rows of 3 is same amount of blocks as 3 rows of 4

$$a \times b \qquad b \times a$$

△ **Two terms**
When multiplying together two terms, the terms can be in any order.

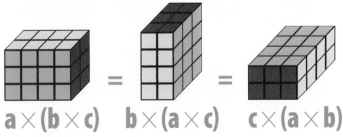

$$a \times (b \times c) \quad b \times (a \times c) \quad c \times (a \times b)$$

△ **Three terms**
Multiplication of three terms can be done in any order.

Sequences

A SEQUENCE IS A SERIES OF NUMBERS WRITTEN AS A LIST THAT FOLLOWS A PARTICULAR PATTERN, OR "RULE".

SEE ALSO

❰ 32–35 Powers and roots
❰ 160–161 What is algebra?
Working with expressions 164–165 ❱
Formulas 169–171 ❱

Each number in a sequence is called a "term". The value of any term in a sequence can be worked out by using the rule for that sequence.

The terms of a sequence

The first number in a sequence is the first term, the second number in a sequence is the second term, and so on.

▷ **A basic sequence**
For this sequence, the rule is that each term is the previous term with 2 added to it.

rule for this sequence is each term equals previous term plus 2

+2   +2   +2   +2

fifth term is 10

first term is 2

dots show sequence continues

2, 4, 6, 8, 10, …

1st term   2nd term   3rd term   4th term   5th term

Finding the "nth" value

The value of a particular term can be found without writing out the entire sequence up until that point by writing the rule as an expression and then using this expression to work out the term.

▷ **The rule as an expression**
Knowing the expression, which is 2n in this example, helps find the value of any term.

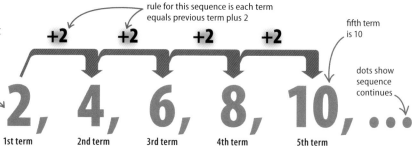

expression used to find value of term – 1 is substituted for n in 1st term, 2 in 2nd term, and so on

2n

means 2 × n — substitute 1 for n

$2n = 2 \times 1 = 2$

1st term

To find the first term, substitute 1 for n.

$2n = 2 \times 2 = 4$

2nd term

To find the second term, substitute 2 for n.

$2n = 2 \times 41 = 82$

41st term

To find the 41st term, substitute 41 for n.

$2n = 2 \times 1{,}000 = 2{,}000$

1,000th term

For the 1,000th term, substitute 1,000 for n. The term here is 2,000.

In the example below, the expression is 4n − 2. Knowing this, the rule can be shown to be: each term is equal to the previous term plus 4.

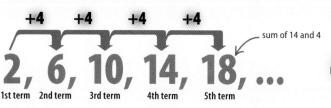

+4   +4   +4   +4

2, 6, 10, 14, 18, …

1st term   2nd term   3rd term   4th term   5th term

sum of 14 and 4

expression here is 4 multiplied by n, minus 2

4n − 2

value of term

$4n - 2 = 4 \times 1 - 2 = 2$

1st term

To find the first term, substitute 1 for n.

$4n - 2 = 4 \times 2 - 2 = 6$

2nd term

To find the second term, substitute 2 for n.

$4n - 2 = (4 \times 1{,}000{,}000) - 2 = 3{,}999{,}998$

1,000,000th term

For the 1,000,000th term, substitute 1,000,000 for n. The term here is 3,999,998.

IMPORTANT SEQUENCES

Some sequences have rules that are slightly more complicated; however, they can be very significant. Two examples of these are square numbers and the Fibonacci sequence.

Square numbers

A square number is found by multiplying a whole number by itself. These numbers can be drawn as squares. Each side is the length of a whole number, which is multiplied by itself to make the square number.

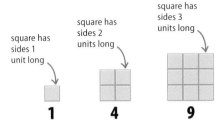

square has sides 1 unit long

square has sides 2 units long

square has sides 3 units long

square has sides 4 units long

square has sides 5 units long

1   **4**   **9**   **16**   **25**

Fibonacci sequence

The Fibonacci sequence is a widely recognized sequence, appearing frequently in nature and architecture. The first two terms of the sequence are both 1, then after this each term is the sum of the two terms that came before it.

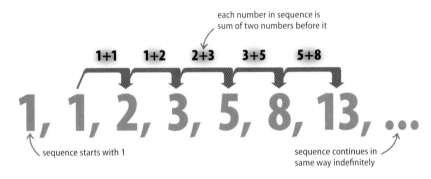

each number in sequence is sum of two numbers before it

1+1   1+2   2+3   3+5   5+8

1, 1, 2, 3, 5, 8, 13, ...

sequence starts with 1

sequence continues in same way indefinitely

Fibonacci and nature

Evidence of the Fibonacci sequence is found everywhere, including in nature. The sequence forms a spiral (see below) and it can be seen in the spiral of a shell (as shown here) or in the arrangement of the seeds in a sunflower. It is named after Leonardo Fibonacci, an Italian mathematician.

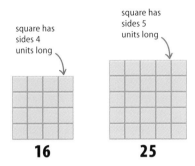

How to draw a Fibonacci spiral

A spiral can be drawn using the numbers in the Fibonacci sequence, by drawing squares with sides as long as each term in the sequence, then drawing curves to touch the opposite corners of these squares.

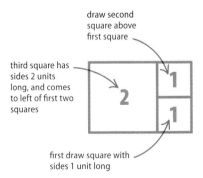

draw second square above first square

third square has sides 2 units long, and comes to left of first two squares

first draw square with sides 1 unit long

First, draw a square that is 1 unit long by 1 unit wide. Draw an identical one above it, then a square with sides 2 units long next to the 1 unit squares. Each square represents a term of the sequence.

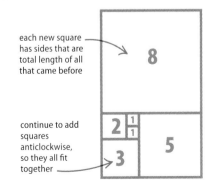

each new square has sides that are total length of all that came before

continue to add squares anticlockwise, so they all fit together

Keep drawing squares that represent the terms of the Fibonacci sequence, adding them in an anticlockwise direction. This diagram shows the first six terms of the sequence.

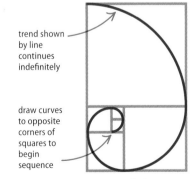

trend shown by line continues indefinitely

draw curves to opposite corners of squares to begin sequence

Finally, draw curves to touch the opposite corners of each square, starting at the centre and working outwards anticlockwise. This curve is a Fibonacci spiral.

2ab Working with expressions

AN EXPRESSION IS A COLLECTION OF SYMBOLS, SUCH AS X AND Y, AND
OPERATIONS, SUCH AS + AND −. IT CAN ALSO CONTAIN NUMBERS.

SEE ALSO

⟨ **160–161** What is algebra?

Formulas **169–171** ⟩

Expressions are important and occur everywhere in mathematics. They can
be simplified to as few parts as possible, making them easier to understand.

Like terms in an expression

Each part of an expression is called a "term". A term can be a number, a symbol, or a number with
a symbol. Terms with the same symbols are "like terms" and it is possible to combine them.

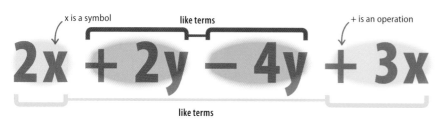

x is a symbol   like terms   + is an operation

$$2x + 2y − 4y + 3x$$

like terms

◁ **Identifying like terms**
The terms 2x and 3x are like terms
because they both contain the symbol
x. The terms 2y and −4y are also like terms
because each contains the symbol y.

Simplifying expressions involving addition and subtraction

When an expression is made up of a number of terms that are to be added or subtracted, there
are a number of important steps that need to be followed in order to simplify it.

▷ **Write down the expression**
Before simplifying the
expression, write it out in a line
from left to right.

$$3a − 5b + 6b − 2a + 3b − 7b$$

▷ **Group the like terms**
Then group the like terms
together, keeping the
operations as they are.

$$3a − 2a \quad − 5b + 6b + 3b − 7b$$

like terms   like terms

▷ **Work out the result**
The next step is to work
out the value of each
like term.

3a − 2a = 1a → $$1a − 3b$$ ← −5b + 6b + 3b − 7b = −3b

▷ **Simplify the result**
Further simplify the result
by removing any 1s in front
of symbols.

term 1a always
written as a → $$a − 3b$$

Simplifying expressions involving multiplication

To simplify an expression that involves terms linked by multiplication signs, the individual numbers and symbols first need to be separated from each other.

simplified expressions are written without multiplication signs

$$6a \times 2b$$

The term 6a means $6 \times a$, and the term 2b means $2 \times b$.

$$6 \times a \times 2 \times b$$

Separate the expression into the individual numbers and symbols involved.

$$12 \times ab = \mathbf{12ab}$$

The product of multiplying 6 and 2 is 12, and that of multiplying a and b is ab. The simplified expression is 12ab.

Simplifying expressions involving division

To simplify an expression involving division, look for any possible cancellation. This means looking to divide all terms of the expression by the same number or letter.

q^2 means $q \times q$

this means $2 \times q$

$$6pq^2 \div 2q$$

same as ÷

$$\frac{6 \times p \times q \times q}{2 \times q}$$

divide 6 by 2, leaving 3

divide q by q, leaving 1

$$\frac{\overset{3}{\cancel{6}} \times p \times q \times \cancel{q}}{\underset{1}{\cancel{2}} \times \cancel{q}_{1}}$$

divide 2 by 2, leaving 1

divide q by q, leaving 1

3pq divided by 1 is simply 3pq

$$\frac{3pq}{1} = \mathbf{3pq}$$

Look for any chances to cancel the expression down and make it smaller and easier to understand. Start by writing the division sum as a fraction.

Both terms in this example are cancelled down by dividing them by 2 and q.

Cancelling by dividing each term equally makes the expression smaller.

Substitution

If the value of each symbol in an expression is known, for example that $y = 2$, the overall value of the expression can be found. This is called "substituting" the values in the expression or "evaluating" the expression.

Substitute the values in the expression $2x - 2y - 4y + 3x$ if

$$x = 1 \text{ and } y = 2$$

L = LENGTH

W = WIDTH

◁ **Substituting values**
The formula for the area of a rectangle is length × width. Substituting 5cm for the length and 8cm for the width, gives an area of 5cm × 8cm = 40cm².

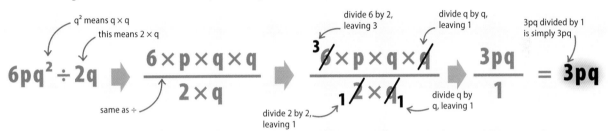

like terms

$$2x - 2y - 4y + 3x$$

like terms

grouped terms are easier to substitute into

$$5x - 6y$$

substitute 1 for x

$$5x = 5 \times 1 = 5$$
$$-6y = -6 \times 2 = -12$$

substitute 2 for y

answer is −7

$$5 - 12 = \mathbf{-7}$$

Group like terms together to simplify the expression.

The expression has now been simplified.

Then substitute the given values for x and y.

The final answer is found to be −7.

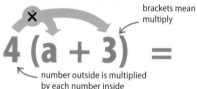

Expanding and factorizing expressions

SEE ALSO

⟨ **164–165** Working with expressions

Quadratic expressions **168** ⟩

THE SAME EXPRESSION CAN BE WRITTEN IN DIFFERENT WAYS – MULTIPLIED OUT (EXPANDED) OR GROUPED INTO ITS COMMON FACTORS (FACTORIZED).

How to expand an expression

The same expression can be written in a variety of ways, depending on how it will be used. Expanding an expression involves multiplying all the parts it contains (terms) and writing it out in full.

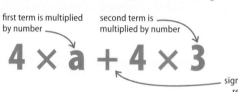

brackets mean multiply

first term is multiplied by number

second term is multiplied by number

$4 \times a = 4a$

$4 \times 3 = 12$

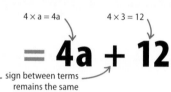

$$4\,(a + 3) = 4 \times a + 4 \times 3 = 4a + 12$$

number outside is multiplied by each number inside

sign between terms remains the same

To expand an expression with a number outside a bracket, multiply all the terms inside the bracket by that number. The bracket means multiply.

▶ **Multiply each term** inside the bracket by the number outside. The sign between the two terms (letters and numbers) remains the same.

▶ **Simplify the resulting terms** to show the expanded expression in its final form. Here, $4 \times a$ is simplified to $4a$ and 4×3 is simplified to 12.

Expanding multiple brackets

To expand an expression that contains two brackets, each part of the first bracket is multiplied by each part of the second bracket. To do this, split up the first (blue) bracket into its parts. Multiply the second (yellow) bracket by the first part and then by the second part of the first bracket.

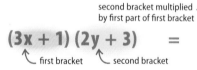

second bracket multiplied by first part of first bracket

$3x \times 2y = 6xy$

$3x \times 3 = 9x$ $1 \times 2y = 2y$ $1 \times 3 = 3$

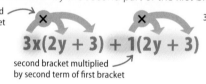 

$$(3x + 1)\,(2y + 3) = 3x(2y + 3) + 1(2y + 3) = 6xy + 9x + 2y + 3$$

first bracket   second bracket

second bracket multiplied by second term of first bracket

these signs remain

To expand an expression of two brackets, multiply all the terms of the second by all the terms of the first.

▶ **Break down the first bracket** into its terms. Multiply the second bracket by each term from the first in turn.

▶ **Simplify the resulting terms** by carrying out each multiplication sum. The signs remain the same.

Squaring a bracket

Squaring a bracket simply means multiplying a bracket by itself. Write it out as two brackets next to each other, and then multiply it to expand as shown above.

multiplying a negative and a positive equals a negative, so $-3 \times x = -3x$

multiplying two negatives equals a positive, so $-3 \times -3 = 9$

$x \times -3 = -3x$

$x \times x = x^2$

$$(x - 3)^2 = (x - 3)\,(x - 3) = x(x - 3) - 3(x - 3) = x^2 - 3x - 3x + 9 = x^2 - 6x + 9$$

multiply second bracket by first part of first bracket

sign remains the same

multiply second bracket by second part of first bracket

To expand a squared bracket, first write the expression out as two brackets next to each other.

▶ **Split the first bracket** into its terms and multiply the second bracket by each term in turn.

▶ **Simplify the resulting terms,** making sure to multiply their signs correctly. Finally, add or subtract like terms (see pp.164–165) together.

How to factorize an expression

Factorizing an expression is the opposite of expanding an expression. To do this, look for a factor (number or letter) that all the terms (parts) of the expression have in common. The common factor can then be placed outside a bracket enclosing what is left of the other terms.

4 is common to both 4b and 12 (because they can both be divided by 4)

this is the same as 12

both b and + 3 are not common to both parts so they go inside the bracket

place 4 outside bracket

remaining factors go inside bracket

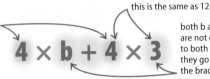

$$4b + 12$$

this means $4 \times b$

$$4 \times b + 4 \times 3$$

$$4(b + 3)$$

bracket means multiply

To factorize an expression, look for any letter or number (factor) that all its parts have in common.

In this case, 4 is a common factor of both 4b and 12, as both can be divided by 4. Divide each part by 4 to find the remaining factors of each part; these go inside the bracket.

Simplify the expression by placing the common factor (4) outside a bracket. The other two factors are placed inside the bracket.

Factorizing more complex expressions

Factorizing can make it simpler to understand and write complex expressions with many terms. Find the factors that all parts of the expression have in common.

3×3

$3 \times 3 \times x \times x \times y = 9x^2y$

$x \times x$

3×5

$3 \times 5 \times x \times y \times y = 15xy^2$

$x \times y^2$

$2 \times 3 \times 3 \times x \times y \times y \times y = 18xy^3$

$y \times y$

$$9x^2y + 15xy^2 + 18xy^3$$

all 3 terms multiplied

To factorize an expression write out the factors of each part, for example, y^2 is $y \times y$. Then look for the numbers and letters that are common to all the factors.

common factor of numbers

common factor of x and x^2

$$3xy$$

common factor of y, y^2, and y^3

All the parts of the expressions contain the letters x and y, and can be factorized by the number 3. These factors are combined to produce one common factor.

3xy is common factor of all parts of the expression

$9x^2y \div 3xy = 3x$

$15xy^2 \div 3xy = 5y$

$18xy^3 \div 3xy = 6y^2$

$$3xy(3x + 5y + 6y^2)$$

Set the common factor (3xy) outside a set of brackets. Inside the brackets, write what remains of each part when divided by it.

LOOKING CLOSER

Factorizing a formula

The formula for finding the surface area (see pp.148–149) of a shape can be worked out using known formulas for the areas of its parts. The formula can look daunting, but it can be made much easier to use by factorizing it.

two circles, one at each end

radius

height

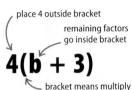

r

h

◁ **Surface of a cylinder**
The formula for the surface area of a cylinder is worked out by adding together the areas of the circles at each end and the rectangle that forms the space between them.

length of rectangle is circumference of circle (2πr)

area of rectangle is length (2πr) × height (h)

area of a circle is πr², for 2 circles it is 2πr²

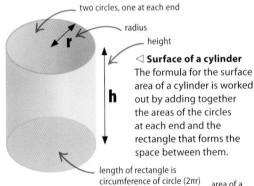

$$2\pi rh + 2\pi r^2$$

To find the formula for the surface area of a cylinder, add together the formulas for the areas of its parts.

2πr is common to both expressions

means multiply by

h and r are not common to both terms so they sit inside the bracket

$$2\pi r (h + r)$$

To make the formula easier to use, simplify it by identifying the common factor, in this case 2πr, and setting it outside the brackets.

Quadratic expressions

SEE ALSO

❮ **166–167** Expanding and factorising expressions

Factorizing quadratic equations **182–183** ❯

A QUADRATIC EXPRESSION CONTAINS AN UNKNOWN TERM (VARIABLE) SQUARED, SUCH AS X^2.

An expression is a collection of mathematical symbols, such as x and y, and operations, such as + and −. A quadratic expression typically contains a squared variable (x^2), a number multiplied by the same variable (x), and a number.

What is a quadratic expression?

A quadratic expression is usually given in the form $ax^2 + bx + c$, where a is the multiple of x^2, b is the multiple of x, and c is the number. a, b, and c can represent any positive or negative numbers.

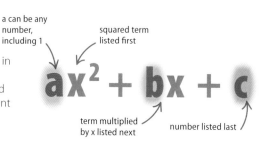

a can be any number, including 1

squared term listed first

term multiplied by x listed next

number listed last

◁ **Quadratic expression**
The standard form of a quadratic expression is one with a squared term (x^2) listed first, terms multiplied by x listed second, and the number listed last.

From two brackets to a quadratic expression

Some quadratic expressions can be factorized to form two expressions within brackets, each containing a variable (x) and an unknown number. Conversely, multiplying out these expressions gives a quadratic expression.

Multiplying two expressions in brackets means multiplying every term of one bracket with every term of the other. The final answer will be a quadratic expression.

To multiply the two brackets, split one of the brackets into its terms. Multiply all the terms of the second bracket first by the x term and then by the numerical term of the first bracket.

Multiplying both terms of the second bracket by each term of the first in turn results in a squared term, two terms multiplied by x, and two numerical terms multiplied together.

Simplify the expression by adding the x terms. This means adding the numbers together inside brackets and multiplying the result by an x outside.

Looking back at the original quadratic expression, it is possible to see that the numerical terms are added to give b, and multiplied to give c.

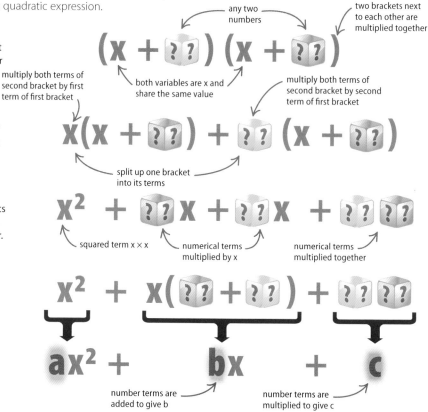

any two numbers

two brackets next to each other are multiplied together

multiply both terms of second bracket by first term of first bracket

both variables are x and share the same value

multiply both terms of second bracket by second term of first bracket

split up one bracket into its terms

squared term x × x

numerical terms multiplied by x

numerical terms multiplied together

number terms are added to give b

number terms are multiplied to give c

 # Formulas

IN MATHS, A FORMULA IS BASICALLY A "RECIPE" FOR FINDING THE VALUE
OF ONE THING (THE SUBJECT) WHEN OTHERS ARE KNOWN.

SEE ALSO

❰ 66–67 Personal finance

❰ 164–165 Working with expressions

Solving equations **172–173 ❱**

A formula usually has a single subject and an equals sign, together with an
expression written in symbols that indicates how to find the subject.

Introducing formulas

The recipe that makes up a formula can be simple
or complicated. However, formulas usually have
three basic parts: a single letter at the
beginning (the subject); an equals sign
that links the subject to the recipe;
and the recipe itself, which
when used, works out the
value of the subject.

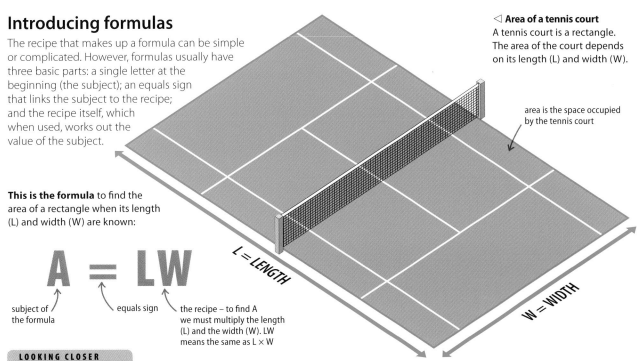

◁ **Area of a tennis court**
A tennis court is a rectangle.
The area of the court depends
on its length (L) and width (W).

area is the space occupied
by the tennis court

L = LENGTH

W = WIDTH

This is the formula to find the
area of a rectangle when its length
(L) and width (W) are known:

$$A = LW$$

subject of
the formula

equals sign

the recipe – to find A
we must multiply the length
(L) and the width (W). LW
means the same as L × W

Formula triangles

Formulas can be rearranged to make
different parts the subject of the
formula. This is useful if the unknown
value to be found is not the subject of
the original formula – the formula can
be rearranged so that the unknown
becomes the subject, making solving
the formula easier.

◁ **Simple rearrangement**
This triangle shows the different
ways the formula for finding a
rectangle can be rearranged.

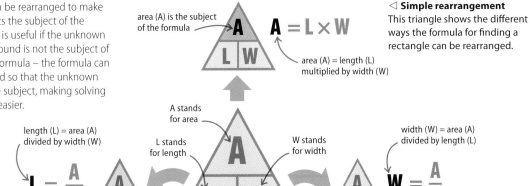

area (A) is the subject
of the formula

$$A = L \times W$$

area (A) = length (L)
multiplied by width (W)

A stands
for area

L stands
for length

W stands
for width

length (L) = area (A)
divided by width (W)

$$L = \frac{A}{W}$$

length (L) is the subject
of the formula

width (W) = area (A)
divided by length (L)

$$W = \frac{A}{L}$$

width (W) is the subject
of the formula

CHANGING THE SUBJECT OF A FORMULA

Changing the subject of a formula involves moving letters or numbers (terms) from one side of the formula to the other, leaving a new term on its own. The way to do this depends on whether the term being moved is positive (+c), negative (-c), or whether it is part of a multiplication (bc) or division (b/c). When moving terms, whatever is done to one side of the formula needs to be done to the other.

Moving a positive term

 −c is brought in to the left of the equals sign

−c is brought in to the right of the equals sign

$$A = b + c$$

$$A - c = b + c - c$$

To make b the subject, +c needs to be moved to the other side of the equals sign.

Add −c to both sides. To move +c, its opposite (−c) must first be added to both sides of the formula to keep it balanced.

+c−c cancels out because c − c = 0

$$A - c = b + \cancel{c} - \cancel{c}$$

Simplify the formula by cancelling out −c and +c on the right, leaving b by itself as the subject of the formula.

a formula must have a single symbol on one side of the equals sign

$$A - c = b$$

The formula can now be rearranged so that it reads **b = A − c.**

Moving a negative term

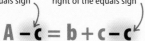 +c is brought in to the left of the equals sign

+c is brought in to the right of the equals sign

$$A = b - c$$

$$A + c = b - c + c$$

To make b the subject, −c needs to be moved to the other side of the equals sign.

Add +c to both sides. To move −c, its opposite (+c) must first be added to both sides of the formula to keep it balanced.

−c+c cancels out because c − c = 0

$$A + c = b - \cancel{c} + \cancel{c}$$

Simplify the formula by cancelling out −c and +c on the right, leaving b by itself as the subject of the formula.

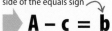 a formula must have a single symbol on one side of the equals sign

$$A + c = b$$

The formula can now be rearranged so that it reads **b = A + c.**

Moving a term in a multiplication sum

bc means b × c

$$A = bc$$

÷c (or /c) is brought in to the left of the equals sign

÷c (or /c) is brought in to the right of the equals sign

$$\frac{A}{c} = \frac{bc}{c}$$

In this example, b is multiplied by c. To make b the subject, ×c needs to move to the other side.

Divide both sides by c. To move ×c, its opposite (÷c) must first be put into both sides of the formula.

c/c cancels out because c/c equals 1

$$\frac{A}{c} = \frac{b\cancel{c}}{\cancel{c}}$$

Simplify the formula by cancelling out c/c on the right, leaving b by itself as the subject of the formula.

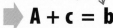 a formula must have a single symbol on one side of the equals sign

$$\frac{A}{c} = b$$

The formula can now be rearranged so that it reads **b = A/c.**

Moving a term in a division sum

 b/c means b ÷ c

$$A = \frac{b}{c}$$

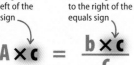 ×c brought in to the left of the equals sign

×c is brought in to the right of the equals sign

$$A \times c = \frac{b \times c}{c}$$

In this example, b is divided by c. To make b the subject, ÷c needs to move to the other side.

Multiply both sides by c. To move ÷c, its opposite (×c) must first be put in to both sides of the formula.

 c/c cancels out because c/c equals 1

$$A \times c = \frac{b\cancel{c}}{\cancel{c}}$$

Simplify the formula by cancelling out c/c on the right, leaving b by itself as the subject of the formula.

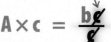 remember that A × c is written as Ac

 a formula must have a single symbol on one side of the equals sign

$$Ac = b$$

The formula can now be rearranged so that it reads **b = Ac.**

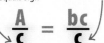

FORMULAS IN ACTION

A formula can be used to calculate how much interest (the amount a bank pays someone in exchange for being able to borrow their money) is paid into a bank account over a particular period of time. The formula for this is principal (or amount of money) × rate of interest × time ÷ 100. This formula is shown here.

this stands for principal, which just means the amount

this stands for rate of interest

$$I = \frac{PRT}{100}$$

this stands for the time it will take to earn interest

this stands for interest

There is a bank account with £/$500 in it, earning simple interest (see pp.66–67) at 2% a year. To find out how much time (T) it will take to earn interest of £/$50, the formula above is used. First, the formula must be rearranged to make T the subject. Then the real values can be put in to work out T.

▷ **Move P**
The first step is to divide each side of the formula by P to move it to the left of the equals sign.

$$I = \frac{PRT}{100} \quad \Rightarrow \quad \frac{I}{P} = \frac{RT}{100}$$

to remove ×P from the right side, divide each side of the formula by P

remember that dividing the right side by P gives **P**RT/**P**100, but the **P**s cancel out, leaving **RT/100**

to remove ×R from the right side, divide each side of the formula by R

▷ **Move R**
The next step is to divide each side of the formula by R to move it to the left of the equals sign.

$$\frac{I}{P} = \frac{RT}{100} \quad \Rightarrow \quad \frac{I}{PR} = \frac{T}{100}$$

remember that dividing the right side by R gives RT/**R**100, but the **R**s cancel out, leaving **T/100**

to remove 100 from the right side, multiply each side of the formula by 100

▷ **Move 100**
Then multiply each side of the formula by 100 to move it to the left of the equals sign.

$$\frac{I}{PR} = \frac{T}{100} \quad \Rightarrow \quad \frac{I\,100}{PR} = T \quad \Rightarrow \quad T = \frac{I\,100}{PR}$$

remember that multiplying the right side by 100 gives **100**T/**100**, but the 100s cancel out, leaving just T

interest (I) is £/$50

length of time (T) to earn interest of £/$50 is 5 years

▷ **Put in real values**
Put in the real values for I (£/$50), P (£/$500), and R (2%) to find the value of T (the time it will take to earn interest of £/$50).

$$T = \frac{I\,100}{PR} \quad \Rightarrow \quad \frac{50 \times 100}{500 \times 2} = \textbf{5 years}$$

principal (P) is £/$500

rate of interest (R) is 2%

Solving equations

AN EQUATION IS A MATHEMATICAL STATEMENT THAT CONTAINS AN EQUALS SIGN.

SEE ALSO

⟨ **160–161** What is algebra?

⟨ **164–165** Working with expressions

⟨ **169–171** Formulas

Linear graphs **174–177** ⟩

Equations can be rearranged to find the value of an unknown variable, such as x or y.

Simple equations

Equations can be rearranged to find the value of an unknown number, or variable. A variable is represented by a letter, such as x or y. Whatever action is taken on one side of an equation must also be made on the other side, so that both sides remain equal.

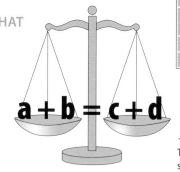

$$a + b = c + d$$

Left-hand side   Right-hand side

◁ **Balancing**
The expressions on either side of the equals sign in an equation are always equal.

To find the value of x the equation must be rearranged so that x is by itself on one side of the equation.

to get rid of this 2, 2 must also be taken from the other side

variable

this expression has the same value as the expression on the other side of the equals sign

$$2 + x = 8$$

Changes made to one side of the equation must also be made to the other side. Subtract 2 from both sides to isolate x.

subtract 2 on this side

as 2 was subtracted from the other side, it must also be subtracted from this side

$$2 + x - 2 = 8 - 2$$

Simplify the equation by cancelling out the +2 and −2 on the left side. This leaves x on its own on the left.

cancel out +2 and −2, which gives 0

$$\cancel{2} + x \cancel{-2} = 8 - 2$$

Once x is the subject of the equation, working out the right side of the equation gives the value of x.

x is now the subject of the equation

$$x = 6$$

working out the right side of the equation (8 − 2) gives the value of x (6)

Creating an equation

Equations can be created to explain day-to-day situations. For example, a taxi firm charges £3 to pick up a customer, and £2 per kilometre travelled. This can be written as an equation.

pick-up cost

cost per kilometre multiplied by distance

$$c = 3 + 2d$$

total cost of the journey

If a customer pays £18 for a journey, the equation can be used to work out how far the customer travelled.

total cost of journey

cost per kilometre multiplied by distance

$$18 = 3 + 2d$$

pick-up cost

Substitute the cost of the journey into the equation.

$$15 = 2d$$

3 has been taken from this side

3 has been taken from this side

Rearrange the equation – subtract 3 from both sides.

$$7\tfrac{1}{2}\,\text{km} = d$$

divide this side by 2

to get rid of 2 in 2d, divide both sides by 2

Find the distance travelled by dividing both sides by 2.

MORE COMPLICATED EQUATIONS

More complicated equations are rearranged in the same way as simple equations – anything done to simplify one side of the equation must also be done to the other side so that both sides of the equation remain equal. The equation will give the same answer no matter where the rearranging is started.

Example 1

This equation has numerical and unknown terms on both sides, so it needs several rearrangements to solve.

numerical term

a appears on both sides of the equation

there are numerical terms on both sides

$$3 + 2a = 5a - 9$$

First, rearrange the numerical terms. To remove the –9 from the right-hand side, add 9 to both sides of the equation.

add 9 to 3

add 9 to –9, which leaves 0 so 5a is isolated

$$12 + 2a = 5a$$

Next, rearrange so that the a's are on the opposite side to the number. This is done by subtracting 2a from both sides.

2a – 2a = 0, leaving 12 on its own

5a –2a = 3a

$$12 = 3a$$

Then rearrange again to make a the only subject of the equation. As the equation contains 3a, divide the whole equation by 3.

the right side must be divided by 3 to isolate a, so the left side must also be divided by 3 to keep both sides equal

divide 3a by 3 to leave a on its own

$$\frac{12}{3} = \frac{3a}{3}$$

The subject of the equation, a, is now on its own on the right side of the equation, and there is only a number on the other side.

12 ÷ 3 = 4, which is the value of a

a is now the subject, isolated by itself on one side of the equation

$$4 = a$$

Reverse the equation to show the unknown variable (a) first. This does not affect the meaning of the equation, as both sides are equal.

put the variable first

this is the solution of the equation – it gives the value of the variable (a)

$$a = 4$$

Example 2

This equation has unknown and numerical terms on both sides, so it will take several rearrangements to solve.

there are numerical values on both sides

there are terms including the unknown a on both sides

$$6a + 4 = 5 - 2a$$

First rearrange the numerical terms. Subtract 4 from both sides of the equation, so that there are numbers on only one side.

4 – 4 = 0, so 6a is on its own

take 4 from 5, leaving 1

$$6a = 1 - 2a$$

Then rearrange the equation so that the unknown variable is on the opposite side to the number, by adding 2a to both sides.

6a + 2a = 8a

–2a + 2a = 0, so 1 is on its own

$$8a = 1$$

Finally, divide each side by 8 to make a the subject of the equation, and to find the solution of the equation.

divide 8a by 8 to leave a by itself on the left of the equation

because the left side was divided by 8 to isolate a, the right side must also be divided by 8 to keep both sides equal

$$a = \frac{1}{8}$$

Linear graphs

GRAPHS ARE A WAY OF PICTURING AN EQUATION. A LINEAR
EQUATION ALWAYS HAS A STRAIGHT LINE.

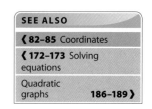
SEE ALSO
⟨ 82–85 Coordinates
⟨ 172–173 Solving equations
Quadratic graphs **186–189 ⟩**

Graphs of linear equations

A linear equation is an equation that does not contain a squared variable such as x^2, or a variable of a higher power, such as x^3. Linear equations can be represented by straight line graphs, where the line passes through coordinates that satisfy the equation. For example, one of the sets of coordinates for $y = x + 5$ is (1, 6), because $6 = 1 + 5$.

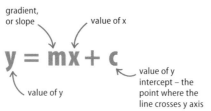

△ **The equation of a straight line**
All straight lines have an equation. The value of m is the gradient (or slope) of the line and c is where it cuts the y axis.

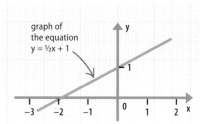

△ **A linear graph**
The graph of an equation is a set of points with coordinates that satisfy the equation.

Finding the equation of a line

To find the equation of a given line, use the graph to find its gradient and y intercept. Then substitute them into the equation for a line, y = mx + c.

To find the gradient of the line (m), draw lines out from a section of the line as shown. Then divide the vertical distance by horizontal distance – the result is the gradient.

$$\text{gradient} = \frac{\text{vertical distance}}{\text{horizontal distance}} \qquad \frac{4}{4} = +1 \leftarrow \text{gradient}$$
division sign

To find the y intercept, look at the graph and find where the line crosses the y axis. This is the y intercept, and is c in the equation.

$$\text{y intercept} = +4$$

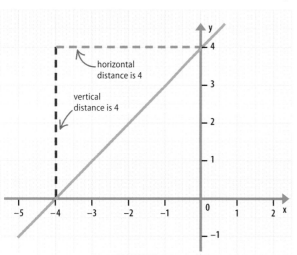

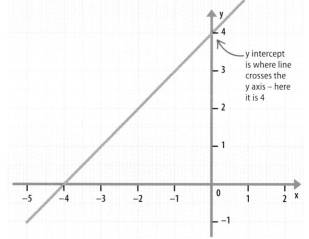

Finally, substitute the values that have been found from the graph into the equation for a line. This gives the equation for the line shown above.

gradient is +1   y intercept is 4   1x simplifies to x

$$y = mx + c \quad \Rightarrow \quad y = x + 4$$

Positive gradients

Lines that slope upwards from left to right have positive gradients. The equation of a line with a positive gradient can be worked out from its graph, as described below.

Find the gradient of the line by choosing a section of it and drawing horizontal (green) and vertical (red) lines out from it so they meet. Count the units each new line covers, then divide the vertical by the horizontal distance.

$$\text{gradient} = \frac{\text{vertical distance}}{\text{horizontal distance}} = \frac{6}{3} = +2$$

+ sign means line slopes upwards from left to right

The y intercept can be easily read off the graph – it is the point where the line crosses the y axis.

$$\text{y intercept} = +1$$

Substitute the values for the gradient and y intercept into the equation of a line to find the equation for this given line.

gradient is +2

y intercept is 1

$$y = \mathbf{mx + c}$$

gradient

y intercept

$$y = \mathbf{2x + 1}$$

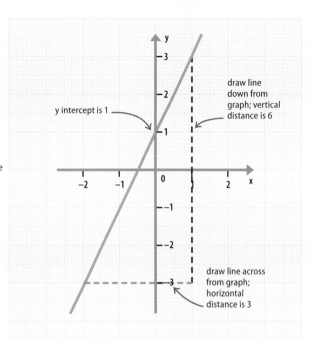

y intercept is 1

draw line down from graph; vertical distance is 6

draw line across from graph; horizontal distance is 3

Negative gradients

Lines that slope downwards from left to right have negative gradients. The equation of these lines can be worked out in the same way as for a line with a positive gradient.

Find the gradient of the line by choosing a section of it and drawing horizontal (green) and vertical (red) lines out from it so they meet. Count the units each new line covers, then divide the vertical by the horizontal distance.

$$\text{gradient} = \frac{\text{vertical distance}}{\text{horizontal distance}} = \frac{4}{1} = 4 \Rightarrow -4$$

insert minus sign to show line slopes downwards from left to right

The y intercept can be easily read off the graph – it is the point where the line crosses the y axis.

$$\text{y intercept} = -4$$

Substitute the values for the gradient and y intercept into the equation of a line to find the equation for this given line.

y intercept is –4

gradient is –4

$$y = \mathbf{mx + c}$$

gradient

y intercept

$$y = \mathbf{-4x - 4}$$

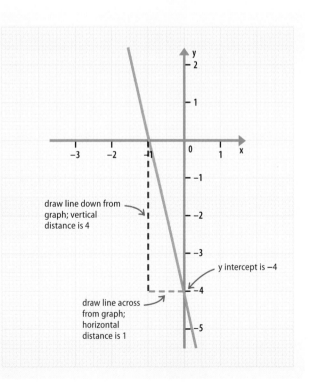

draw line down from graph; vertical distance is 4

y intercept is –4

draw line across from graph; horizontal distance is 1

How to plot a linear graph

The graph of a linear equation can be drawn by working out several different sets of values for x and y and then plotting these values on a pair of axes. The x values are measured along the x axis, and the y values along the y axis.

this means 2 multiplied by x

▷ **The equation**
This shows that each of the y values for this equation will be double the size of each of the x values.

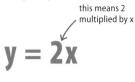

$$y = 2x$$

first, choose some possible values of x

x	y = 2x
1	2
2	4
3	6
4	8

then find corresponding values of y by doubling each x value

First, choose some possible values of x, numbers below 10 are easiest to work with. Find the corresponding values of y using a table. Put the x values in the first column, then multiply each number by 2 to find the corresponding values for y.

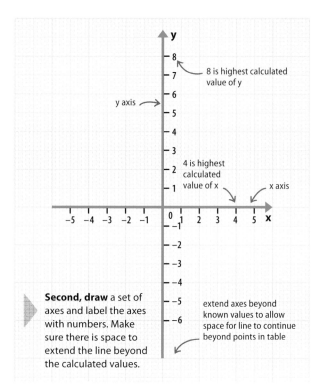

8 is highest calculated value of y

y axis →

4 is highest calculated value of x

x axis

▷ **Second, draw** a set of axes and label the axes with numbers. Make sure there is space to extend the line beyond the calculated values.

extend axes beyond known values to allow space for line to continue beyond points in table

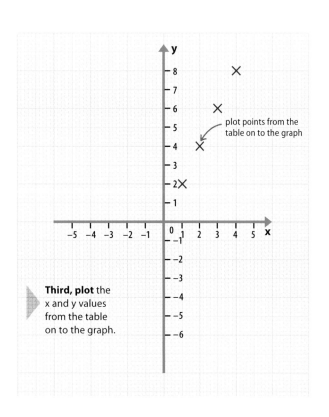

plot points from the table on to the graph

▷ **Third, plot** the x and y values from the table on to the graph.

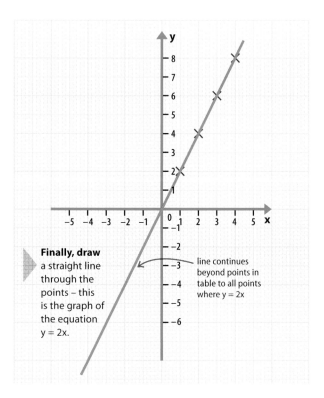

▷ **Finally, draw** a straight line through the points – this is the graph of the equation y = 2x.

line continues beyond points in table to all points where y = 2x

Downward-sloping graph

Graphs of linear equations can slope downwards or upwards from left to right. Downward-sloping graphs have a negative gradient; upward-sloping graphs have a positive gradient.

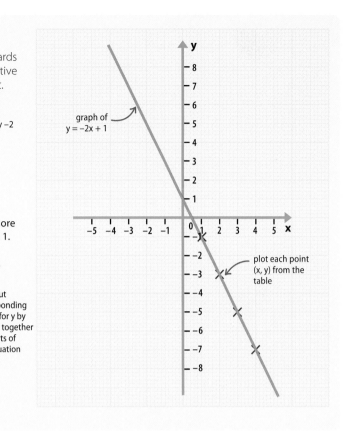

graph of
$y = -2x + 1$

plot each point (x, y) from the table

The equation here contains the term –2x. Because x is multiplied by a negative number (–2), the graph will slope downwards.

this means x multiplied by –2

$$y = -2x + 1$$

Use a table to find some values for x and y. This equation is more complex than the last, so add more rows to the table: –2x and 1. Calculate each of these values, then add them to find y. It is important to keep track of negative signs in front of numbers.

write down some possible values of x

x	–2x	+1	y=–2x+1
1	–2	+1	–1
2	–4	+1	–3
3	–6	+1	–5
4	–8	+1	–7

work out corresponding values for y by adding together the parts of the equation

values of x multiplied by –2

+1 is constant

Temperature conversion graph

A linear graph can be used to show the conversion between the two main methods of measuring temperature - Fahrenheit and Celsius. To convert any temperature from Fahrenheit into Celsius, start at the position of the Fahrenheit temperature on the y axis, read horizontally across to the line, and then vertically down to the x axis to find the Celsius value.

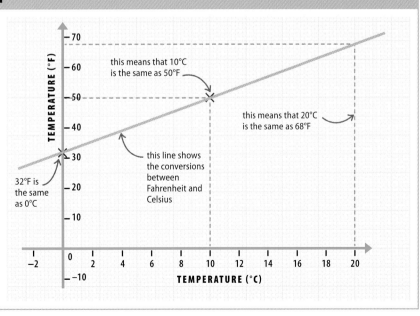

this means that 10°C is the same as 50°F

this means that 20°C is the same as 68°F

this line shows the conversions between Fahrenheit and Celsius

32°F is the same as 0°C

°F	°C
32.0	0
50.0	10

△ **Temperature conversion**
Two sets of values for Fahrenheit (F) and Celsius (C) give all the information that is needed to plot the conversion graph.

Simultaneous equations

SEE ALSO

❮ **164–165** Working with expressions

❮ **169–171** Formulas

SIMULTANEOUS EQUATIONS ARE PAIRS OF EQUATIONS WITH THE SAME UNKNOWN VARIABLES, THAT ARE SOLVED TOGETHER.

Solving simultaneous equations

Simultaneous equations are pairs of equations that contain the same variables and are solved together. There are three ways to solve a pair of simultaneous equations: elimination, substitution, and by graph; they all give the same answer.

both equations contain the variable x

both equations contain the variable y

$$3x - 5y = 4$$
$$4x + 5y = 17$$

◁ **A pair of equations**
These simultaneous equations both contain the unknown variables x and y.

Solving by elimination

Make the x or y terms the same for both equations, then add or subtract them to eliminate that variable. The resulting equation finds the value of one variable, which is then used to find the other.

▷ **Equation pair**
Solve this pair of simultaneous equations using the elimination method.

$$10x + 3y = 2$$
$$2x + 2y = 6$$

Multiply or divide one of the equations to make one variable the same as in the other equation. Here, the second equation is multiplied by 5 to make the x terms the same.

the second equation is multiplied by 5

$$10x + 3y = 2$$
first equation stays as it is

$$2x + 2y = 6 \quad \xrightarrow{\times 5} \quad 10x + 10y = 30$$

this is the second equation

the x term is now the same as in the first equation

the second equation is multiplied by 5, so both equations now have the same value of x (10x)

Then add or subtract each set of terms in the second equation from or to each set in the first, to remove the matching terms. The new equation can then be solved. Here, the second equation is subtracted from the first, and the remaining variables are rearranged to isolate y.

this will cancel out the x terms

$$10x - 10x + 3y - 10y = 2 - 30$$

subtract the numerical terms from each other as well as the unknown terms

the x terms have been eliminated as 10x − 10x = 0

$$-7y = -28$$

this side is divided by −7 to isolate y

$$y = \frac{-28}{-7}$$

this side must also be divided by −7

$$y = 4$$
this gives the value of y

Choose one of the two original equations – it does not matter which – and put in the value for y that has just been found. This eliminates the y variable from the equation, leaving only the x variable. Rearranging the equation means that it can be solved, and the value of the x can be found.

$$2x + 2y = 6$$
the second equation has been chosen

$$2x + (2 \times 4) = 6$$
it is already known that y = 4 so 2y = 8

$$2x + 8 = 6$$
2 × 4 = 8

subtracting 8 from this side to isolate 2x

$$2x = -2$$
subtract 8 from this side: 6 − 8 = −2

divide this side by 2 to isolate x

$$\frac{2x}{2} = \frac{-2}{2}$$
this side must also be divided by 2

$$x = -1$$
this is the value of x

Both unknown variables have now been found – these are the solutions to the original pair of equations.

$$x = -1 \qquad y = 4$$

Solving by substitution

To use this method, rearrange one of the two equations so that the two unknown values (variables) are on different sides of the equation, then substitute this rearranged equation into the other equation. The new, combined equation contains only one unknown value and can be solved. Substituting the new value into one of the equations means that the other variable can also be found. Equations that cannot be solved by elimination can usually be solved by substitution.

▷ **Equation pair**
Solve this pair of simultaneous equations using the substitution method.

$$x + 2y = 7$$
$$4x - 3y = 6$$

Choose one of the equations, and rearrange it so that one of the two unknown values is the subject. Here x is made the subject by subtracting 2y from both sides of the equation.

choose one of the equations; this is the first equation

$$x + 2y = 7$$

make x the subject by subtracting 2y from both sides of the equation

$$x = 7 - 2y$$

2y must be subtracted from both sides of the equation

Then substitute the expression that has been found for that variable (x = 7 −2y) into the other equation. This gives only one unknown value in the newly compiled equation. Rearrange this new equation to isolate y and find its value.

substitute the expression for x which has been found in the previous step

take the other equation

$$4x - 3y = 6$$

$$4(7 - 2y) - 3y = 6$$

this equation now has only one unknown value so it can be solved

$$28 - 8y - 3y = 6$$

multiply out the brackets above: 4 × 7 = 28 and 4 × −2y = −8y

$$28 - 11y = 6$$

simplify the two y terms: −8y −3y = −11y

$$-11y = -22$$

isolate the y term by subtracting 28 from this side

28 must also be subtracted from this side: 6 − 28 = −22

$$\frac{-11y}{-11} = \frac{-22}{-11}$$

divide this side by −11 to isolate y (−11y ÷ −11 = y)

this side must also be divided by −11

$$y = 2$$

this is the value of y

Substitute the value of y that has just been found into either of the original pair of equations. Rearrange this equation to isolate x and find its value.

choose one of the equations; this is the first one

$$x + 2y = 7$$

$$x + (2 \times 2) = 7$$

seeing as y = 2, 2y is 2 × 2 = 4

$$x + 4 = 7$$

work out the terms in the brackets: 2 × 2 = 4

$$x = 3$$

as 4 has been subtracted from the other side of the equation, it must also be subtracted from this side: 7 − 4 = 3

subtract 4 from this side to isolate x

Both unknown variables have now been found – these are the solutions to the original pair of equations.

$$x = 3 \qquad y = 2$$

Solving simultaneous equations with graphs

Simultaneous equations can be solved by rearranging each equation so that it is expressed in terms of y, using a table to find sets of x and y coordinates for each equation, then plotting the graphs. The solution is the coordinates of the point where the graphs intersect.

▷ **A pair of equations**
This pair of simultaneous equations can be solved using a graph. Each equation will be represented by a line on the graph.

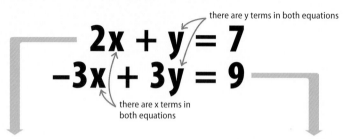

there are y terms in both equations

$$2x + y = 7$$
$$-3x + 3y = 9$$

there are x terms in both equations

To isolate y in the first equation, rearrange the equation so that y is left on its own on one side of the equals sign. Here, this is done by subtracting 2x from both sides of the equation.

2x + y = 7 is the first equation

$$2x + y = 7$$

−2x has also been added to this side

−2x has been added to this side to cancel out the 2x and isolate y

$$y = 7 - 2x$$

To isolate y in the second equation, rearrange the equation so that y is left on its own on one side of the equals sign. Here, this is done by first adding 3x to both sides, then dividing both sides by 3.

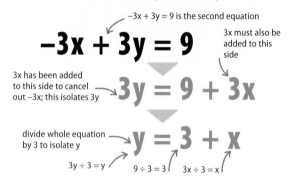

−3x + 3y = 9 is the second equation

$$-3x + 3y = 9$$

3x must also be added to this side

3x has been added to this side to cancel out −3x; this isolates 3y

$$3y = 9 + 3x$$

divide whole equation by 3 to isolate y

$$y = 3 + x$$

3y ÷ 3 = y   9 ÷ 3 = 3   3x ÷ 3 = x

Find the corresponding x and y values for the rearranged first equation using a table. Choose a set of x values that are close to zero, then use the table to work out the y values.

Find the corresponding x and y values for the rearranged second equation using a table. Choose the same set of x values as for the other table, then use the table to work out the y values.

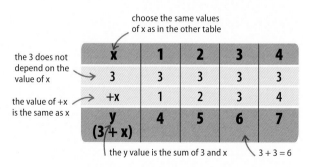

the 7 does not depend on x

choose a set of values for x that are close to 0

x	1	2	3	4
7	7	7	7	7
−2x	−2	−4	−6	−8
y (7 − 2x)	5	3	1	−1

work out the value of −2x for each value of x

the y value is the sum of 7 and −2x

7 − 6 = 1

choose the same values of x as in the other table

the 3 does not depend on the value of x

the value of +x is the same as x

x	1	2	3	4
3	3	3	3	3
+x	1	2	3	4
y (3 + x)	4	5	6	7

the y value is the sum of 3 and x

3 + 3 = 6

Draw a set of axes, then plot the two sets of x and y values. Join each set of points with a straight line, continuing the line past where the points lie. If the pair of simultaneous equations has a solution, then the two lines will cross.

Unsolvable simultaneous equations

Sometimes a pair of simultaneous equations does not have a solution. For example, the graphs of the two equations $x + y = 1$ and $x + y = 2$ are always equidistant from each other (parallel) and, because they do not intersect, there is no solution to this pair of equations.

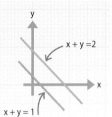

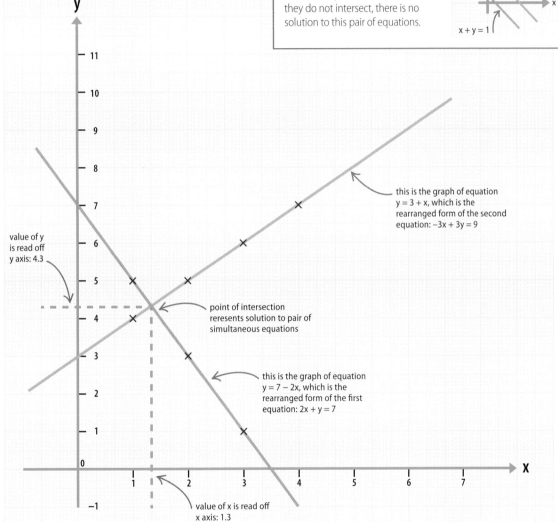

value of y is read off y axis: 4.3

point of intersection reresents solution to pair of simultaneous equations

this is the graph of equation $y = 3 + x$, which is the rearranged form of the second equation: $-3x + 3y = 9$

this is the graph of equation $y = 7 - 2x$, which is the rearranged form of the first equation: $2x + y = 7$

value of x is read off x axis: 1.3

The solution to the pair of simultaneous equations is the coordinates of the point where the two lines cross. Read from this point down to the x axis and across to the y axis to find the values of the solution.

$$x = 1.3 \qquad y = 4.3$$

Factorizing quadratic equations

SOME QUADRATIC EQUATIONS (EQUATIONS IN THE FORM AX² + BX + C = 0) CAN BE SOLVED BY FACTORIZING.

Quadratic factorization

Factorization is the process of finding the terms that multiply together to form another term. A quadratic equation is factorized by rearranging it into two bracketed parts, each containing a variable and a number. To find the values in the brackets, use the rules from multiplying brackets (see p.168) – that the numbers add together to give b and multiply together to give c of the original quadratic equation.

SEE ALSO

❮ **168** Quadratic expressions

❮ **184–185** The quadratic formula

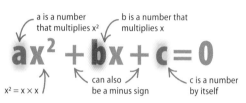

brackets set next to each other are multiplied together

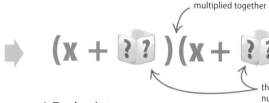

these two unknown numbers add together to give b and multiply together to give c of the original equation

a is a number that multiplies x²

b is a number that multiplies x

$$ax^2 + bx + c = 0$$

x² = x × x

can also be a minus sign

c is a number by itself

△ **A quadratic equation**
All quadratic equations have a squared term (x²), a term that is multiplied by x, and a numerical term. The letters a, b, and, c all stand for different numbers.

△ **Two brackets**
A quadratic equation can be factorized as two brackets, each containing an x and a number. Multiplied out, they result in the equation.

Solving simple quadratic equations

To solve quadratic equations by factorization, first find the missing numerical terms in the brackets. Then solve each bracket separately to find the answers to the original equation.

To solve a quadratic equation, first look at its b and c terms. The terms in the two brackets will need to add together to give b (6 in this case) and multiply together to give c (8 in this case).

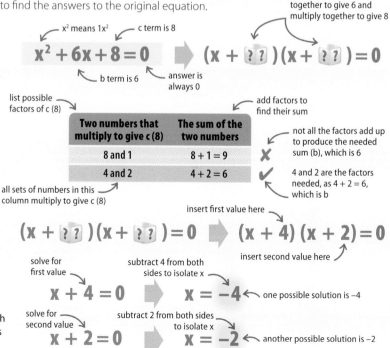

x² means 1x²   c term is 8

these two numbers add together to give 6 and multiply together to give 8

$$x^2 + 6x + 8 = 0$$

b term is 6   answer is always 0

$$(x + \boxed{??})(x + \boxed{??}) = 0$$

To find the unknown terms, draw a table. In the first column, list the possible combinations of numbers that multiply together to give the value of c (8). In the second column, add these terms together to see if they add up to b (6).

list possible factors of c (8)

add factors to find their sum

Two numbers that multiply to give c (8)	The sum of the two numbers
8 and 1	8 + 1 = 9
4 and 2	4 + 2 = 6

all sets of numbers in this column multiply to give c (8)

✗ not all the factors add up to produce the needed sum (b), which is 6

✓ 4 and 2 are the factors needed, as 4 + 2 = 6, which is b

Insert the factors into the brackets after the x terms. Because the two brackets multiplied together are equal to the original quadratic expression, they can also be set to equal 0.

insert first value here

$$(x + \boxed{??})(x + \boxed{??}) = 0 \implies (x + 4)(x + 2) = 0$$

insert second value here

For the two brackets to multiply to equal 0, the value of either bracket needs to be 0. Set each bracket equal to 0 and solve. The resulting values are the two solutions of the original equation.

solve for first value

subtract 4 from both sides to isolate x

$$x + 4 = 0 \implies x = -4$$

one possible solution is −4

solve for second value

subtract 2 from both sides to isolate x

$$x + 2 = 0 \implies x = -2$$

another possible solution is −2

Solving more complex quadratic equations

Quadratic equations do not always appear in the standard form of $ax^2 + bx + c = 0$. Instead, several x^2 terms, x terms, and numbers may appear on both sides of the equals sign. However, if all terms appear at least once, the equation can be rearranged in the standard form, and solved using the same methods as for simple equations.

these terms need to be moved to other side of equation for it to equal 0

This equation is not written in standard quadratic form, but contains an x^2 term and a term multiplied by x so it is known to be one. In order to solve, it needs to be rearranged to equal 0.

$$x^2 + 11x + 13 = 2x - 7$$

7 has been added to this side (13 + 7 = 20)

7 has been added to this side, which cancels out −7, leaving 2x on its own

Start by moving the numerical term from the right-hand side of the equals sign to the left by adding its opposite to both sides of the equation. In this case, −7 is moved by adding 7 to both sides.

$$x^2 + 11x + 20 = 2x$$

adding −2x to 11x gives 9x

subtracting 2x from this side cancels out 2x

Next, move the term multiplied by x to the left of the equals sign by adding its opposite to both sides of the equation. In this case, 2x is moved by subtracting 2x from both sides.

$$x^2 + 9x + 20 = 0$$

list possible factors of c (20)

add the factors to find their sum

It is now possible to solve the equation by factorizing. Draw a table for the possible numerical values of x. In one column, list all values that multiply together to give the c term, 20; in the other, add them together to see if they give the b term (9).

Factors of +20	Sum of factors	
20, 1	21	✗
2, 10	12	✗
5, 4	9	✔

stop when the factors add to the b term, 9

all sets of numbers in this column multiply to give 20

brackets set next to each other are multiplied together

entire equation equals 0

Write the correct pair of factors into brackets and set them equal to 0. The two factors of the quadratic (x + 5) and (x + 4) multiply together to give 0, therefore one of the factors must be equal to 0.

$$(x + 5)(x + 4) = 0$$

subtract 5 from both sides to isolate x

solve for first value

$$x + 5 = 0 \quad\blacktriangleright\quad x = -5$$

one possible solution is −5

Solve the quadratic equation by solving each of the bracketed expressions separately. Make each bracketed expression equal to 0, then find its solution. The two resulting values are the two solutions to the quadratic equation: −5 and −4.

solve for second value

subtract 4 from both sides to isolate x

$$x + 4 = 0 \quad\blacktriangleright\quad x = -4$$

another possible solution is −4

LOOKING CLOSER

Not all quadratic equations can be factorized

Some quadratic equations cannot be factorized, as the sum of the factors of the purely numerical component (c term) does not equal the term multiplied by x (b term). These equations must be solved by formula (see pp.184–185).

b term (3)

c term (1)

$$x^2 + 3x + 1 = 0$$

both sets of numbers multiply together to give c (1)

Factors of +1	Sum of factors	
1, 1	2	✗
−1, −1	−2	✗

a sum of +3 is needed as the b term is 3

The equation above is a typical quadratic equation, but cannot be solved by factorizing.

Listing all the possible factors and their sums in a table shows that there is no set of factors that add to b (3), and multiply to give c (1).

SEE ALSO

‹ 169-171 Formulas

‹ 182-183 Factorizing quadratic equations

Quadratic graphs   186-189 ›

x^2 The quadratic formula

QUADRATIC EQUATIONS CAN BE SOLVED USING A FORMULA.

The quadratic formula

The quadratic formula can be used to solve any quadratic equation. Quadratic equations take the form $ax^2 + bx + c = 0$, where a, b, and c are numbers and x is the unknown.

△ **A quadratic equation**
Quadratic equations include a number multiplied by x^2, a number multiplied by x and a number by itself.

number that multiplies x^2
number that multiplies x
number with no x terms

$$ax^2 + bx + c = 0$$

△ **The quadratic formula**
The quadratic formula allows any quadratic equation to be solved. Substitute the different values in the equation into the quadratic formula to solve the equation.

$$x = \frac{-b \pm \sqrt{b^2 - 4ac}}{2a}$$

this means add or subtract

Quadratic variations

Quadratic equations are not always the same. They can include negative terms, or terms with no numbers in front of them ("x" is the same as "1x") and do not always equal 0.

the values in the equation can be negative as well as positive

quadratic equations are not always equal to 0

$$-4x^2 + x - 3 = 8$$

when an x appears without a number in front of it, x=1

Using the quadratic formula

To use the quadratic formula, substitute the values for a, b, and c in a given equation into the formula, then work through the formula to find the answers. Take great care with the signs (+, –) of a, b, and c.

Given a quadratic equation, work out the values of a, b, and c. Once these values are known, substitute them into the quadratic formula, making sure that their positive and negative signs do not change. In this example, a is 1, b is 3, and c is –2.

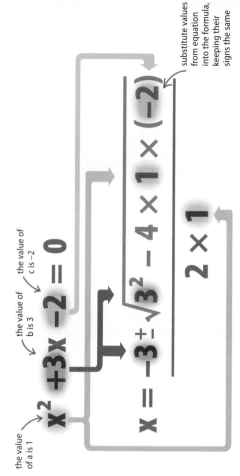

the value of a is 1

the value of b is 3

the value of c is –2

$$x^2 + 3x - 2 = 0$$

$$x = \frac{-3 \pm \sqrt{3^2 - 4 \times 1 \times (-2)}}{2 \times 1}$$

substitute values from equation into the formula, keeping their signs the same

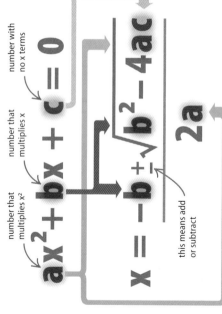

Work through the formula step-by-step to find the answer to the equation. First simplify the values under the square root sign. Work out the square of 3 (which equals 9), then work out the value of $4 \times 1 \times -2$ (which equals -8).

$$x = \frac{-3 \pm \sqrt{9 - (-8)}}{2}$$

$3 \times 3 = 9$

$4 \times 1 \times (-2) = -8$

two minus signs cancel out, so $9 - (-8) = 9 + 8$

Work out the numbers under the square root sign: $9 - (-8)$ equals $9 + 8$, which equals 17. Then, use a calculator to find the square root of 17.

$$x = \frac{-3 \pm \sqrt{17}}{2}$$

$9 + 8 = 17$

Once the sum is simplified, it must be split to find the two answers – one when the second value is subtracted from the first, and the other where they are added.

$$x = \frac{-3 \pm 4.12}{2}$$

$-$

$+$

4.12 is the square root of 17 rounded to 2 decimal places

Add the two values on the top part of the fraction; here the values are -3 and 4.12.

$$x = \frac{-3 + 4.12}{2}$$

Divide the top part of the fraction by the bottom part to find an answer.

$$x = \frac{1.12}{2}$$

$-3 + 4.12 = 1.12$

Give both the answers, as quadratic equations always have two solutions.

$$x = 0.56$$

quadratic equations always have two solutions

Subtract the second value on the top part of the fraction from the first value; here the values are -3 and -4.12.

$$x = \frac{-3 - 4.12}{2}$$

Divide the top part of the fraction by the bottom part to find an answer.

$$x = \frac{-7.12}{2}$$

$-3 - 4.12 = -7.12$

$$x = -3.56$$

Quadratic graphs

THE GRAPH OF A QUADRATIC EQUATION IS A SMOOTH CURVE.

SEE ALSO

❰ **30–31** Positive and negative numbers

❰ **168** Quadratic expressions

❰ **174–177** Linear graphs

❰ **182–183** Factorizing quadratic equations

❰ **184–185** The quadratic formula

The exact shape of the curve of a quadratic graph varies, depending on the values of the numbers a, b, and c in the quadratic equation $y = ax^2 + bx + c$.

Quadratic equations all have the same general form: $y = ax^2 + bx + c$. With a particular quadratic equation, the values of a, b, and c are known, and corresponding sets of values for x and y can be worked out and put in a table. These values of x and y are then plotted as points (x,y) on a graph. The points are then joined by a smooth line to create the graph of the equation.

A quadratic equation can be shown as a graph. Pairs of x and y values are needed to plot the graph. In quadratic equations, the y values are given in terms of x – in this example each y value is equal to the value of x squared (x multiplied by itself), added to 3 times x, added to 2.

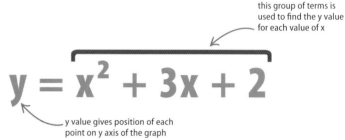

this group of terms is used to find the y value for each value of x

$$y = x^2 + 3x + 2$$

y value gives position of each point on y axis of the graph

Find sets of values for x and y in order to plot the graph. First, choose a set of x values. Then, for each x value, work out the different values (x^2, 3x, 2) for each value at each stage of the equation. Finally, add the stages to find the corresponding y value for each x value.

choose some values of x around 0 — $y = x^2 + 3x + 2$, so it is difficult to work out the y values straight away — work out x^2 in this column — work out 3x in this column — +2 is the same for each x value — add values in each mauve row to find y values

x	y
−3	
−2	
−1	
0	
1	
2	
3	

x	x^2	3x	+2	y
−3	9	−9	2	
−2	4	−6	2	
−1	1	−3	2	
0	0	0	2	
1	1	3	2	
2	4	6	2	
3	9	9	2	

x	x^2	3x	+2	y
−3	9	−9	2	2
−2	4	−6	2	0
−1	1	−3	2	0
0	0	0	2	2
1	1	3	2	6
2	4	6	2	12
3	9	9	2	20

y is the sum of numbers in each mauve row

+ + =

△ **Values of x**
The value of y depends on the value of x, so choose a set of x values and then find the corresponding values of y. Choose x values either side of 0 as they are easiest to work with.

△ **Different parts of the equation**
Each quadratic equation has 3 different parts – a squared x value, a multiplied x value, and an ordinary number. Work out the different values of each part of the equation for each value of x, being careful to pay attention to when the numbers are positive or negative.

△ **Corresponding values of y**
Add the three parts of the equation together to find the corresponding values of y for each x value, making sure to pay attention to when the different parts of the equation are positive or negative.

Draw the graph of the equation. Use the values of x and y that have been found in the table as the coordinates of points on the graph. For example, x = 1 has the corresponding value y = 6. This becomes the point on the graph with the coordinates (1, 6).

▷ **Draw the axes and plot the points**
Draw the axes of the graph so that they cover the values found in the tables. It is often useful to make the axes a bit longer than needed, in case extra values are added later. Then plot the corresponding values of x and y as points on the graph.

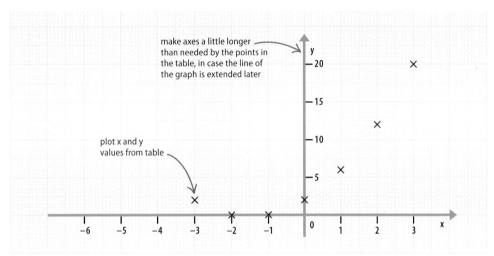

make axes a little longer than needed by the points in the table, in case the line of the graph is extended later

plot x and y values from table

▷ **Join the points**
Draw a smooth line to join the points plotted on the graph. This line is the graph of the equation y = x² + 3x + 2. Bigger and smaller values of x could have been chosen, and so the line continues past the values that have been plotted.

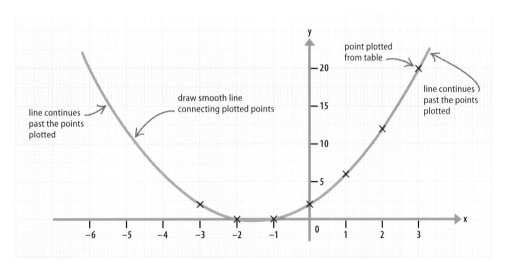

point plotted from table

line continues past the points plotted

draw smooth line connecting plotted points

line continues past the points plotted

The shape of a quadratic graph

The shape of a quadratic graph depends on whether the number which multiplies x² is positive or negative. If it is positive, the graph is a smile; if it is negative, the graph is a frown.

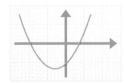

◁ **y = ax² + bx + c**
If the value of the a term is positive, then the graph of the equation is shaped like this.

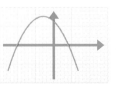

◁ **y = −ax² + bx + c**
If the value of the a term is negative, then the graph of the equation is shaped like this.

Using graphs to solve quadratic equations

A quadratic equation can be solved by drawing a graph. If a quadratic equation has a y value that is not 0, it can be solved by drawing both a quadratic and a linear graph (the linear graph is of the y value that is not 0) and finding where the two graphs cross. The solutions to the equation are the x values where the two graphs cross.

This equation has two parts: a quadratic equation on the left and a linear equation on the right. To find the solutions to this equation, draw the quadratic and linear graphs on the same axes. To draw the graphs, it is necessary to find sets of x and y values for both sides of the equation.

linear part of equation

$$-x^2 - 2x + 3 = -5$$

quadratic part of equation

y values for quadratic part of equation are dependent on value of x

y values for linear part of equation are all −5

$$y = -x^2 - 2x + 3$$

$$y = -5$$

◁ **y = −5**
This graph is very simple: whatever value x takes, y is always −5. This means that the graph is a straight horizontal line that passes through the y axis at −5.

Find values of x and y for the quadratic part of the equation using a table. Choose x values either side of 0 and split the equation into parts (−x², −2x, and +3). Work out the value of each part for each value of x, then add the values of all three parts to find the y value for each x value.

choose some values of x around 0

x	y
−4	
−3	
−2	
−1	
0	
1	
2	

y = −x² − 2x + 3, so it is difficult to work out the y values straight away

work out x² first then put a minus sign in front to give values

work out −2x in this column

+3 is the same for each x value

x	−x²	−2x	3	y
−4	−16	+8	+3	
−3	−9	+6	+3	
−2	−4	+4	+3	
−1	−1	+2	+3	
0	0	0	+3	
1	−1	−2	+3	
2	−4	−4	+3	

add values in each mauve row to find y values

x	−x²	−2x	3	y
−4	−16	+8	+3	−5
−3	−9	+6	+3	0
−2	−4	+4	+3	3
−1	−1	+2	+3	4
0	0	0	+3	3
1	−1	−2	+3	0
2	−4	−4	+3	−5

y is the sum of numbers in each mauve row

+ + =

△ **Values of x**
Each value of y depends on the value of x. Choose a number of values for x, and work out the corresponding values of y. It is easiest to choose values of x that are either side of 0.

△ **Different parts of the equation**
The equation has 3 different parts: −x², −2x, and +3. Work out the values of each part of the equation for each value of x, being careful to pay attention to whether the values are positive or negative. The last part of the equation, +3, is the same for each x value.

△ **Corresponding values of y**
Finally, add the three parts of the equation together to find the corresponding values of y for each x value. Make sure to pay attention to whether the different parts of the equation are positive or negative.

Plot the quadratic graph. First draw a set of axes, then plot the points of the graph, using the values of x and y from the table as the coordinates of each point. For example, when x = −4, y has the value y = −5. This gives the coordinates of the point (−4, −5) on the graph. After plotting the points, draw a smooth line to join them.

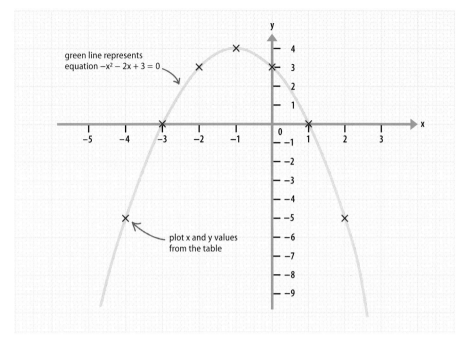

green line represents equation $-x^2 - 2x + 3 = 0$

plot x and y values from the table

Then plot the linear graph. The linear graph (y = −5) is a horizontal straight line that passes through the y axis at −5. The points at which the two lines cross are the solutions to the equation $-x^2 -2x +3 = -5$.

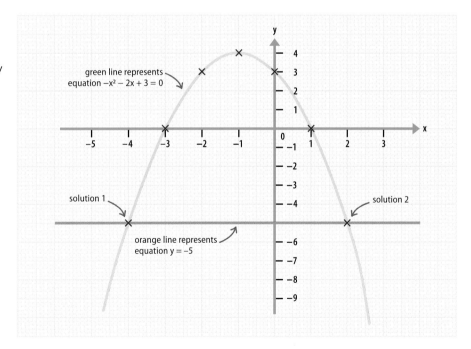

green line represents equation $-x^2 - 2x + 3 = 0$

solution 1

solution 2

orange line represents equation y = −5

The solutions are read off the graph – they are the two x values of the points where the lines cross: −4 and 2.

coordinates of first solution

coordinates of second solution

first solution to the equation

second solution to the equation

$(-4, -5)$ and $(2, -5)$ ➡ $x = -4$ $x = 2$

≠ Inequalities

AN INEQUALITY IS USED TO SHOW THAT ONE
QUANTITY IS NOT EQUAL TO ANOTHER.

SEE ALSO

❮ **30–31** Positive and
negative numbers

❮ **164–165** Working with
expressions

❮ **172–173** Solving
equations

Inequality symbols

An inequality symbol shows that the numbers on either side of
it are different in size and how they are different. There are five
main inequality symbols. One simply shows that two numbers
are not equal, the others show in what way they are not equal.

$$x \neq y$$

◁ **Not equal to**
This sign shows that x
is not equal to y; for
example, 3 ≠ 4.

$$x > y$$

△ **Greater than**
This sign shows that x is greater
than y; for example, 7 > 5.

$$x \geq y$$

△ **Greater than or equal to**
This sign shows that x is greater
than or equal to y.

$$x < y$$

△ **Less than**
This sign shows that x is less
than y. For example, –2 < 1.

$$x \leq y$$

△ **Less than or equal to**
This sign shows that x is less
than or equal to y.

▽ **Inequality number line**
Inequalities can be shown on a number line. The empty circles represent
greater than (>) or less than (<), and the filled circles represent greater
than or equal to (≥) or less than or equal to (≤).

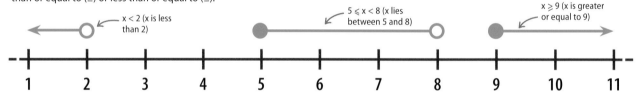

x < 2 (x is less
than 2)

5 ≤ x < 8 (x lies
between 5 and 8)

x ≥ 9 (x is greater
or equal to 9)

LOOKING CLOSER

Rules for inequalities

Inequalities can be rearranged,
as long as any changes are made
to both sides of the inequality. If
an inequality is multiplied or
divided by a negative number,
then its sign is reversed.

▷ **Multiplying or dividing
by a positive number**
When an inequality is multiplied
or divided by a positive number,
its sign does not change.

a ≥ 4

× +3 → 3a ≥ 12
sign stays
the same

÷ +4 → $\frac{a}{4}$ ≥ 1

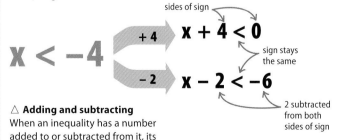

x < –4

+4 → x + 4 < 0
sign stays
the same

–2 → x – 2 < –6

4 added to both
sides of sign

2 subtracted
from both
sides of sign

△ **Adding and subtracting**
When an inequality has a number
added to or subtracted from it, its
sign does not change.

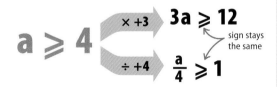

p < 3

× –3 → –3p > –9
sign is
reversed

÷ –1 → –p > –3

△ **Multiplying or dividing by a negative number**
When an inequality is multiplied or divided by a
negative number, its sign is reversed. In this example,
a less than sign becomes a greater than sign.

Solving inequalities

Inequalities can be solved by rearranging them, but anything that is done to one side of the inequality must also be done to the other. For example, any number added to cancel a numerical term from one side must be added to the numerical term on the other side.

To solve this inequality will mean adding 2 to both sides then dividing by 3.

$$3b - 2 \geqslant 10$$

To isolate 3b, −2 needs to be removed, which means adding +2 to both sides.

adding 2 to 3b − 2 leaves 3b on its own → 10 + 2 = 12

$$3b \geqslant 12$$

Solve the inequality by dividing both sides by 3 to isolate b.

3b divided by 3 leaves b on its own

$$b \geqslant 4$$ ← 12 ÷ 3 = 4

To solve this inequality will mean subtracting 3 from both sides then dividing by 3.

$$3a + 3 < 12$$

Rearrange the inequality by subtracting 3 from each side to isolate the a term on the left.

subtracting 3 leaves 3a on its own → 12 − 3 = 9

$$3a < 9$$

Solve the inequality by dividing both sides by 3 to isolate a. This is the solution to the inequality.

3a divided by 3 leaves a on its own → 9 ÷ 3 = 3

$$a < 3$$

Solving double inequalities

To solve a double inequality, deal with each side separately to simplify it, then combine the two sides back together again in a single answer.

This is a double inequality that needs to be split into its two parts for the solution to be found.

$$-1 \leqslant 3x + 5 < 11$$

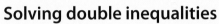

$$-1 \leqslant 3x + 5$$

These are the two parts the double inequality is split into; each one needs to be solved separately.

$$3x + 5 < 11$$

subtracting 5 from −1 gives −6 ↘

subtracting 5 from 3x + 5 leaves 3x on its own

$$-6 \leqslant 3x$$

Isolate the x terms by subtracting 5 from both sides of the smaller parts.

subtracting 5 from 3x + 5 leaves 3x on its own

subtracting 5 from 11 gives 6

$$3x < 6$$

−6 ÷ 3 = −2 ↘   3x ÷ 3 = x

$$-2 \leqslant x$$

Solve the part inequalities by dividing both of them by 3.

3x ÷ 3 = x   6 ÷ 3 = 2

$$x < 2$$

$$-2 \leqslant x < 2$$

Finally, combine the two small inequalities back into a single double inequality, with each in the same position as it was in the original double inequality.

Statistics

What is statistics?

STATISTICS IS THE COLLECTION, ORGANIZATION, AND
PROCESSING OF DATA.

**Organizing and analysing data helps make large quantities of
information easier to understand. Graphs and other visual charts
present information in a way that is instantly understandable.**

Working with data

Data is information, and it is everywhere, in enormous quantities. When data is
collected, for example from a questionnaire, it often forms long lists that are hard
to understand. It can be made easier to understand if the data is reorganized into
tables, and even more accessible by taking the table and plotting its information
as a graph or pie chart. Graphs show trends clearly, making the data much easier
to analyse. Pie charts present data in an instantly accessible way, allowing the
relative sizes of groups to be seen immediately.

group	number
Female teachers	10
Male teachers	5
Female students	66
Male students	19
Total people	100

△ **Collecting data**
Once data has been collected, it must
be organized into groups before it can be
effectively analysed. A table is the usual
way to do this. This table shows the
different groups of people in a school.

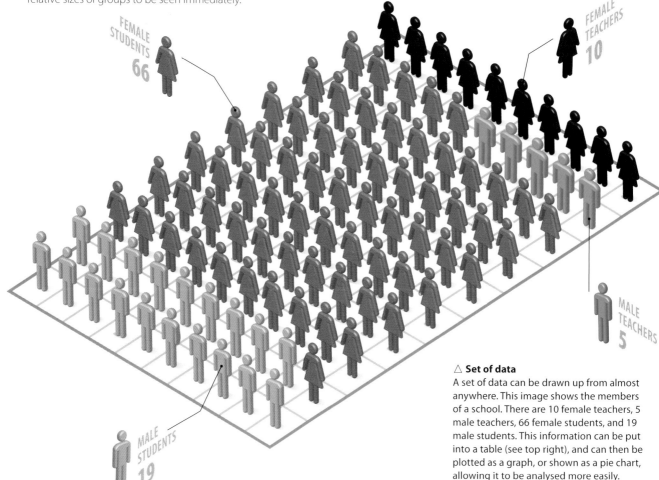

FEMALE
STUDENTS
66

FEMALE
TEACHERS
10

MALE
TEACHERS
5

MALE
STUDENTS
19

△ **Set of data**
A set of data can be drawn up from almost
anywhere. This image shows the members
of a school. There are 10 female teachers, 5
male teachers, 66 female students, and 19
male students. This information can be put
into a table (see top right), and can then be
plotted as a graph, or shown as a pie chart,
allowing it to be analysed more easily.

Presenting data

There are many ways of presenting statistical data. It can be presented simply as a table, or in visual form, as a graph or diagram. Bar charts, pictograms, line graphs, pie charts, and histograms are among the most common ways of showing data visually.

number of times a value appears

Group of data	Frequency
Group 1	4
Group 2	8
Group 3	6
Group 4	4
Group 5	5

△ **Table of data**
Information is put into tables to organize it into categories, to give a better idea of what trends the data shows. The table can then be used to draw a graph, pictogram, or pie chart.

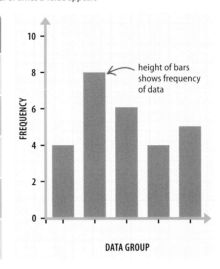

height of bars shows frequency of data

△ **Bar chart**
Bar charts show groups of data on the x axis, and frequency on the y axis. The height of each "bar" shows what frequency of data there is in each group.

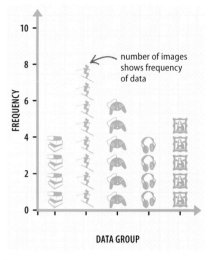

number of images shows frequency of data

△ **Pictogram**
Pictograms are a very basic type of bar chart. Each image on a pictogram represents a number of pieces of information, for example, it could represent four musicians.

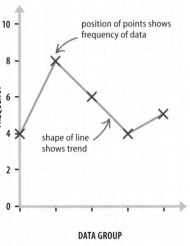

position of points shows frequency of data

shape of line shows trend

△ **Line graph**
Line graphs show data groups on the x axis, and frequency on the y axis. Points are plotted to show the frequency for each group, and lines between the points show trends.

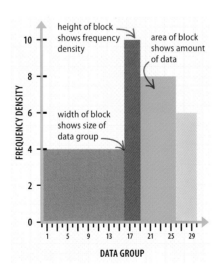

height of block shows frequency density

area of block shows amount of data

width of block shows size of data group

△ **Histogram**
Histograms use the area of rectangular blocks to show the different sizes of groups of data. They are useful for showing data from groups of different sizes.

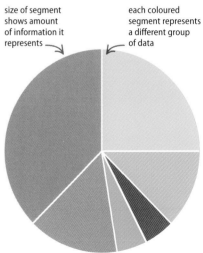

size of segment shows amount of information it represents

each coloured segment represents a different group of data

△ **Pie chart**
Pie charts show groups of information as sections of a circle. The bigger the section of the circle, the larger the amount of data it represents.

Collecting and organizing data

BEFORE INFORMATION CAN BE PRESENTED AND ANALYSED, THE DATA
MUST BE CAREFULLY COLLECTED AND ORGANIZED.

SEE ALSO	
Bar charts	**198–201)**
Pie charts	**202–203)**
Line graphs	**204–205)**

What is data?

In statistics, the information that is collected, usually in the form of lists of numbers, is known as data. To make sense of these lists, the data needs to be sorted into groups and presented in an easy-to-read form, for example as tables or diagrams. Before it is organized, it is sometimes called raw data.

choice of drinks

COLA, ORANGE JUICE,

PINEAPPLE JUICE, MILK,

APPLE JUICE, WATER

◁ **Questions**
Before designing a questionnaire start with an idea of a question to collect data, for example, which soft drinks do children prefer?

Collecting data

A common way of collecting information is in a survey. A selection of people are asked about their preferences, habits, or opinions, often in the form of a questionnaire. The answers they give, which is the raw data, can then be organized into tables and diagrams.

Drink questionnaire

This questionnaire is being used to find out what children's favourite soft drinks are. Put a cross in the box that relates to you.

1) Are you a boy or a girl?

X boy ☐ girl

2) What is your favourite drink?

☐ pineapple juice ☐ orange juice X apple juice

☐ milk ☐ cola ☐ other

information from these answers is collected as lists of data

3) How often do you drink it?

☐ once a week or less X 2–3 times a week ☐ 3–5 times a week

☐ over 5 times a week

▷ **Questionnaire**
Questionnaires often take the form of a series of multiple choice questions. The replies to each question are then easy to sort into groups of data. In this example, the data would be grouped by the soft drinks chosen.

4) Where is you favourite drink usually bought from?

☐ supermarket X convenience store ☐ other

Tallying

Results from a survey can be organized into a chart. The left-hand column shows the groups of data from the questionnaire. A simple way to record the results is by making a tally mark in the chart for each answer. To tally, mark a line for each unit and cross through the lines when 5 is reached.

making tally marks in groups of five makes chart easier to read; the line that goes across is the 5th

Soft drink	Tally
Cola	ЖЖ I
Orange juice	ЖЖ ЖЖ I
Apple juice	II
Pineapple juice	I
Milk	II
Other	I

△ **Tally chart**
This tally chart shows the results of the survey with tally marks.

Soft drink	Tally	Frequency
Cola	ЖЖ I	6
Orange juice	ЖЖ ЖЖ I	11
Apple juice	II	2
Pineapple juice	I	1
Milk	II	2
Other	I	1

△ **Frequency table**
Counting the tally marks for each group, the results (frequency) can be entered in a separate column to make a frequency table.

Tables

Tables showing the frequency of results for each group are a useful way of presenting data. Values from the frequency column can be analysed and used to make charts or graphs of the data. Frequency tables can have more columns to show more detailed information.

Soft drink	Frequency
Cola	6
Orange juice	11
Apple juice	2
Pineapple juice	1
Milk	2
Other	1

△ **Frequency table**
Data can be presented in a table. In this example, the number of children that chose each type of drink is shown.

Soft drink	Boy	Girl	Total
Cola	4	2	6
Orange juice	5	6	11
Apple juice	0	2	2
Pineapple juice	1	0	1
Milk	1	1	2
Other	1	0	1

△ **Two-way table**
This table has extra columns that break down the information further. It also shows the numbers of boys and girls and their preferences.

Bias

In surveys it is important to question a wide selection of people, so that the answers provide an accurate picture. If the survey is too narrow, it may be unrepresentative and show a bias towards a particular answer.

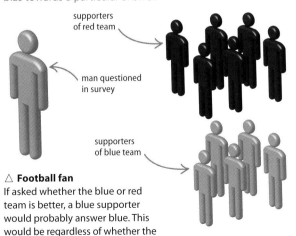

supporters of red team

man questioned in survey

supporters of blue team

△ **Football fan**
If asked whether the blue or red team is better, a blue supporter would probably answer blue. This would be regardless of whether the reds had proved their superiority.

LOOKING CLOSER
Data logging

A lot of data is recorded by machines – information about the weather, traffic, or internet usage for instance. The data can then be organized and presented in charts, tables and graphs that make it easier to understand and analyse.

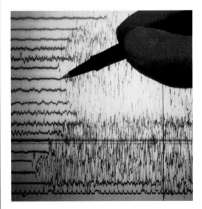

◁ **Seismometer**
A seismometer records movements of the ground that are associated with earthquakes. The collected data is analysed to find patterns that may predict future earthquakes.

Bar charts

BAR CHARTS ARE A WAY OF PRESENTING DATA AS A DIAGRAM.

A bar chart displays a set of data graphically. Bars of different lengths are drawn to show the size (frequency) of each group of data in the set.

SEE ALSO

❮ **196–197** Collecting and organizing data

Pie charts	**202–203** ❯
Line graphs	**204–205** ❯
Histograms	**216–217** ❯

Using bar charts

Presenting data in the form of a diagram makes it easier to read than a list or table. A bar chart shows a set of data as a series of bars, with each bar representing a group within the set. The height of each bar represents the size of each group – a value known as the group's "frequency". Information can be seen clearly and quickly from the height of the bars, and accurate values for the data can be read from the vertical axis of the chart. A bar chart can be drawn with a pencil, a ruler, and graph paper, using information from a frequency table.

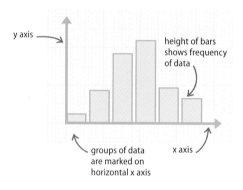

y axis

height of bars shows frequency of data

groups of data are marked on horizontal x axis

x axis

◁ **A bar chart**
In a bar chart, each bar represents a group of data from a particular data set. The size (frequency) of each data group is shown by the height of the corresponding bar.

This frequency table shows the groups of data and the size (frequency) of each group in a data set.

To draw a bar graph, first choose a suitable scale for your data. Then draw a vertical line for the y axis and a horizontal line for the x axis. Label each axis according to the columns of the table, and mark with the data from the table.

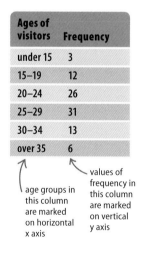

Ages of visitors	Frequency
under 15	3
15–19	12
20–24	26
25–29	31
30–34	13
over 35	6

age groups in this column are marked on horizontal x axis

values of frequency in this column are marked on vertical y axis

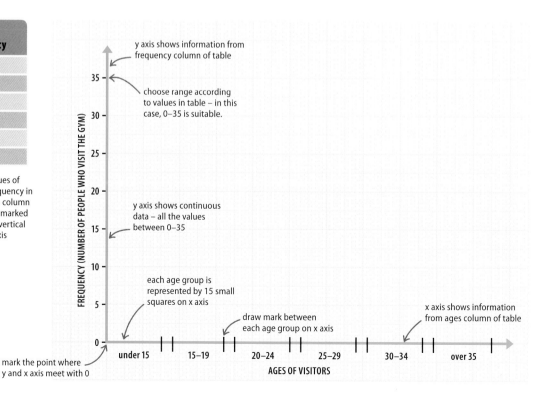

y axis shows information from frequency column of table

choose range according to values in table – in this case, 0–35 is suitable.

y axis shows continuous data – all the values between 0–35

each age group is represented by 15 small squares on x axis

draw mark between each age group on x axis

x axis shows information from ages column of table

FREQUENCY (NUMBER OF PEOPLE WHO VISIT THE GYM)

mark the point where y and x axis meet with 0

under 15   15–19   20–24   25–29   30–34   over 35

AGES OF VISITORS

From the table, take the number (frequency) for the first group of data (3 in this case) and find this value on the vertical y axis. Draw a horizontal line between the value on the y axis and the end of the first age range, marked on the x axis. Next, draw a line for the second frequency (in this case, 12) above the second age group marked on the x axis, and similar lines for all the remaining data.

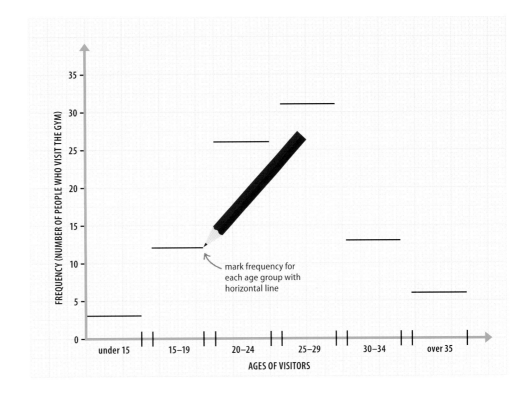

mark frequency for each age group with horizontal line

To complete the bar chart, draw vertical lines up from the dividing marks on the x axis. These will meet the ends of the lines you have drawn from the frequency table, making the bars. Colouring in the bars makes the chart easier to read.

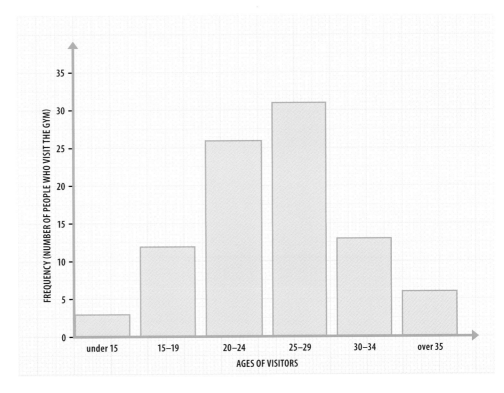

Different types of bar chart

There are several different ways of presenting information in a bar chart. The bars may be drawn horizontally, as three-dimensional blocks, or in groups of two. In every type, the size of the bar shows the size (frequency) of each group of data.

Hobby	Frequency (number of children)
Reading	25
Sport	45
Computer games	30
Music	19
Collecting	15

◁ **Table of data**
This data table shows the results of a survey in which a number of children were asked about their hobbies.

▷ **Horizontal bar chart**
In a horizontal bar chart, the bars are drawn horizontally rather than vertically. Values for the number of children in each group, the frequency, can be read on the horizontal x axis.

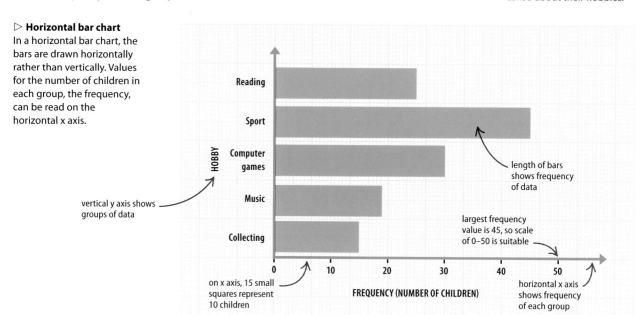

vertical y axis shows groups of data

length of bars shows frequency of data

largest frequency value is 45, so scale of 0–50 is suitable

on x axis, 15 small squares represent 10 children

horizontal x axis shows frequency of each group

▷ **Three-dimensional bar chart**
The three-dimensional blocks in this type of bar chart give it more visual impact, but can be misleading. Because of the perspective, the tops of the blocks appear to show two values for frequency – the true value is read from the front edge of the block.

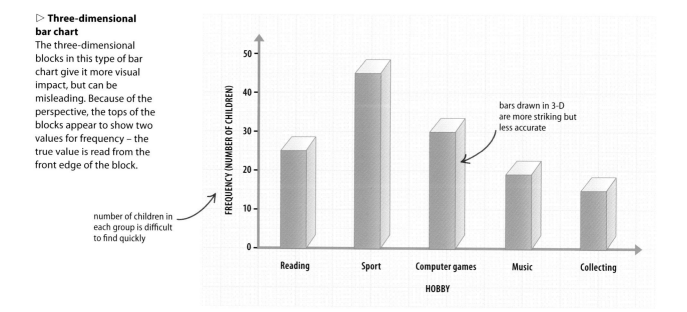

number of children in each group is difficult to find quickly

bars drawn in 3-D are more striking but less accurate

Compound and composite bar charts

For data divided into sub-groups, compound or composite bar charts can be used. In a compound bar chart, bars for each sub-group of data are drawn side by side. In a composite bar chart, two sub-groups are combined into one bar.

Hobby	Boys	Girls	Total frequency
Reading	10	15	25
Sport	25	20	45
Computer games	20	10	30
Music	10	9	19
Collecting	5	10	15

◁ **Table of data**
This data table shows the results of the survey on children's hobbies divided into separate figures for boys and girls.

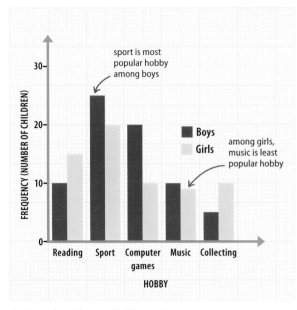

△ **Compound (or multiple) bar chart**
In a compound bar chart, each data group has two or more bars of different colours, that each represent a sub-group of that data. A key shows which colour represents which groups.

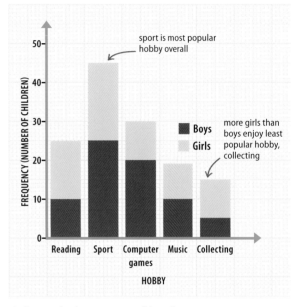

△ **Composite (or component) bar chart**
In a composite bar chart, two or more sub-groups of data are shown as one bar, one sub-group on top of the other. This has the advantage of also showing the total value of the group of data.

Frequency polygons

Another way of presenting the same information as a bar chart is in a frequency polygon. Instead of bars, the data is shown as a line on the chart. The line connects the midpoints of each group of data.

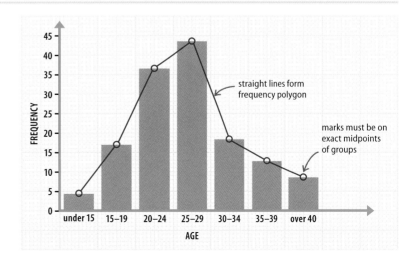

▷ **Drawing a frequency polygon**
Mark the frequency value at the midpoint of each group of data, in this case, the middle of each age range. Join the marks with straight lines.

Pie charts

PIE CHARTS ARE A USEFUL VISUAL WAY TO PRESENT DATA.

A pie chart shows data as a circle divided into segments, or slices, with each slice representing a different part of the data.

SEE ALSO

❬ **76–77** Angles
❬ **142–143** Arcs and sectors
❬ **196–197** Collecting and organizing data
❬ **198–201** Bar charts

Why use a pie chart?

Pie charts are often used to present data as they have an immediate visual impact. The size of each slice of the pie clearly shows the relative sizes of different groups of data, which means that the comparison of data is quick and easy.

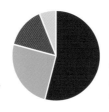

◁ **Reading a pie chart**
When a pie chart is divided into slices, it is easy to understand the information. It is clear in this example that the red section represents the largest group of data.

Identifying data

To get the information necessary to calculate the size, or angle, of each slice of a pie chart, a table of data known as a frequency table is created. This identifies the different groups of data, and shows both their size (frequency of data) and the size of all of the groups of data together (total frequency).

Country of origin	Frequency of data
United Kingdom	375
United States	250
Australia	125
Canada	50
China	50
Unknown	150
TOTAL FREQUENCY	1,000

◁ **Frequency table**
The table shows the number of hits on a website, split into the countries where they occurred.

"frequency of data" is broken down country by country

data from each country is used to calculate size of each slice

"total frequency" is total number of website hits from all countries

▽ **Calculating the angles**
To find the angle for each slice of the pie chart, take the information in the frequency table and use it in this formula.

$$\text{angle} = \frac{\text{frequency of data}}{\text{total frequency}} \times 360°$$

For example:

number of website hits

divide both numbers

angle for pie chart

$$\text{angle for United Kingdom} = \frac{375}{1,000} \times 360° = 135°$$

total number of website hits

The angles for the remaining slices are calculated in the same way, taking the data for each country from the frequency table and using the formula. The angles of all the slices of the pie should add up to 360° – the total number of degrees in a circle.

$$\text{United States} = \frac{250}{1,000} \times 360 = 90°$$

$$\text{Australia} = \frac{125}{1,000} \times 360 = 45°$$

$$\text{Canada} = \frac{50}{1,000} \times 360 = 18°$$

$$\text{China} = \frac{50}{1,000} \times 360 = 18°$$

$$\text{Unknown} = \frac{150}{1,000} \times 360 = 54°$$

United Kingdom

135°

Drawing a pie chart

Drawing a pie chart requires a compass to draw the circle, a protractor to measure the angles accurately, and a ruler to draw the slices of the pie.

First, draw a circle using a compass (see pp.74–75).

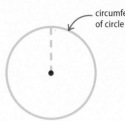

circumference of circle

Draw a straight line from the centre point of the circle to the circumference edge of the circle.

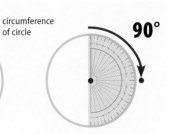

90°

Measure the angle of a slice from the centre and straight line. Mark it on the edge of the circle. Draw a line from the centre to this mark.

centre point

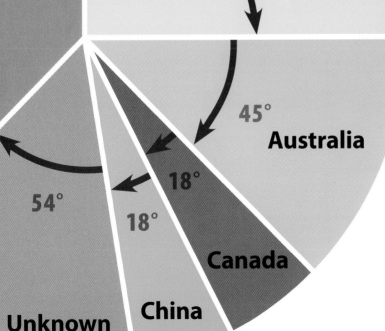

United States
90°

45°
Australia

54°

18°

18°

Canada

Unknown

China

◁ **Finished pie chart**
After drawing each slice on the circle, the pie chart can be labelled and colour coded, as necessary. As the angles add up to 360°, all of the slices fit into the circle exactly.

Labelling pie charts

There are three different ways to label the different slices of a pie chart: with annotation (a, b), with labels (c, d), or with a key (e, f). Annotation and keys can be useful tools when slices are too small to label the required data.

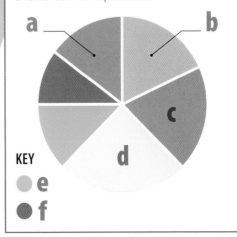

a   b

c

d

KEY
● e
● f

Line graphs

LINE GRAPHS SHOW DATA AS LINES ON A SET OF AXES.

SEE ALSO

❮ **174–177** Linear graphs

❮ **196–197** Collecting and organizing data

Line graphs are a way of accurately presenting information in an easy-to-read form. They are particularly useful for showing data over a period of time.

Drawing a line graph

A pencil, a ruler, and graph paper are all that is needed to draw a line graph. Data from a table is plotted on the graph, and these points are joined to create a line.

Day	Sunshine (hours)
Monday	12
Tuesday	9
Wednesday	10
Thursday	4
Friday	5
Saturday	8
Sunday	11

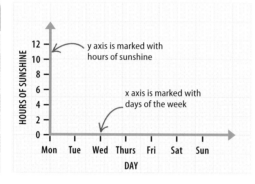

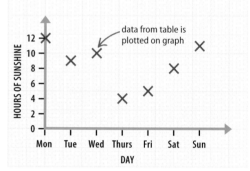

The columns of the table provide the information for the horizontal and vertical lines – the x and y axes.

Draw a set of axes. Label the x axis with data from the first column of the table (days). Label the y axis with data from the second (hours of sunshine).

Read up the y axis from Monday on the x axis and mark the first value. Do this for each day, reading up from the x axis and across from the y axis.

Use a ruler and a pen or pencil to connect the points and complete the line graph once all the data has been marked (or plotted). The resulting line clearly shows the relationship between the two sets of data.

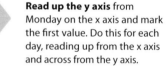

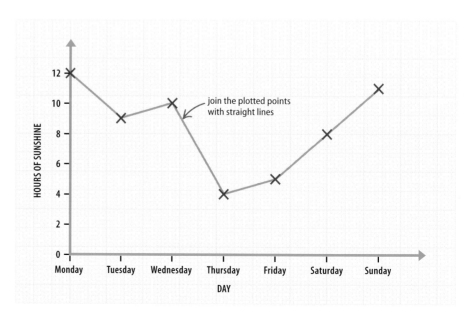

Interpreting line graphs

This graph shows temperature changes over a 24-hour period. The temperature at any time in the day can be found by locating that time on the x axis, reading up to the line, and then across to the y axis.

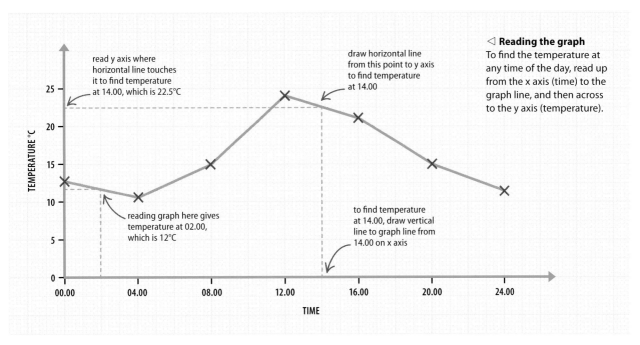

read y axis where horizontal line touches it to find temperature at 14.00, which is 22.5°C

reading graph here gives temperature at 02.00, which is 12°C

draw horizontal line from this point to y axis to find temperature at 14.00

to find temperature at 14.00, draw vertical line to graph line from 14.00 on x axis

◁ **Reading the graph**
To find the temperature at any time of the day, read up from the x axis (time) to the graph line, and then across to the y axis (temperature).

TEMPERATURE °C

TIME

Cumulative frequency graphs

A cumulative frequency diagram is a type of line graph that shows how often each value occurs in a group of data. Joining the points of a cumulative frequency graph with straight lines usually creates an "S" shape, and the curve of the S shows which values occur most frequently within the set of data.

weight is shown in grouped data

frequency, in this case number of people, is shown for each group

cumulative frequency is sum of all frequencies

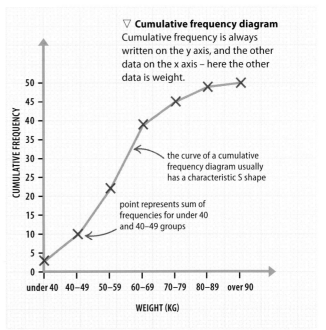

▽ **Cumulative frequency diagram**
Cumulative frequency is always written on the y axis, and the other data on the x axis – here the other data is weight.

Weight (kg)	Frequency	Cumulative frequency
under 40	3	3
40–49	7	10 (3+7)
50–59	12	22 (3+7+12)
60–69	17	39 (3+7+10+17)
70–79	6	45 (3+7+10+17+6)
80–89	4	49 (3+7+10+17+6+4)
over 90	1	50 (3+7+10+17+6+4+1)

◁ **Cumulative frequency**
The frequency is cumulative because each frequency is added to all the frequencies that come before it.

CUMULATIVE FREQUENCY

the curve of a cumulative frequency diagram usually has a characteristic S shape

point represents sum of frequencies for under 40 and 40–49 groups

WEIGHT (KG)

cumulative frequency is plotted on graph

Averages

AN AVERAGE IS A "MIDDLE" VALUE OF A SET OF DATA. IT IS A
TYPICAL VALUE THAT REPRESENTS THE ENTIRE SET OF DATA.

SEE ALSO

❰ **196–197** Collecting and organizing data

Moving averages **210–211 ❱**

Measuring spread **212–215 ❱**

Different types of averages

There are several different types of average. The main ones are called the mean, the median and the mode. Each one gives slightly different information about the data. In everyday life, the term "average" usually refers to the mean.

150, 160, 170, 180, 180

working out averages often requires listing a set of data arranged in ascending order

The mode

The mode is the value that appears most frequently in a set of data. It is easier to find the mode if you put the data list into an ascending order of values (from lowest to highest). If different values appear the same number of times, there may be more than one mode.

this colour represents mode because it appears most often

◁ **The mode colour**
The set of data in this example is a series of coloured figures. The pink people appear the most often, so pink is the mode value.

150, 160, 170, **180, 180**

180 occurs twice in this list, more often than any other value, so it is the mode, or most frequent, value

▷ **Average heights**
The heights of this group of people can be arranged as a list of data. From this list, the different types of average can be found – mean, median, and mode.

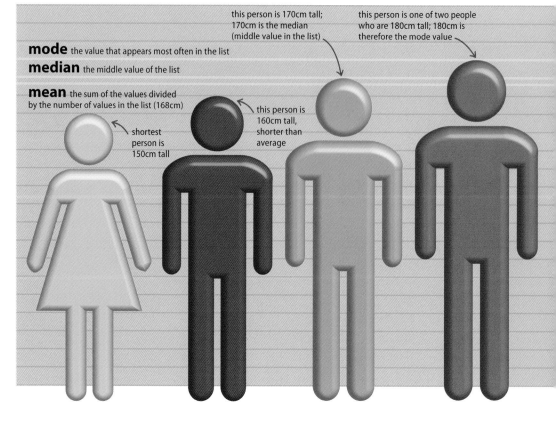

mode the value that appears most often in the list
median the middle value of the list
mean the sum of the values divided by the number of values in the list (168cm)

this person is 170cm tall; 170cm is the median (middle value in the list)

this person is one of two people who are 180cm tall; 180cm is therefore the mode value

this person is 160cm tall, shorter than average

shortest person is 150cm tall

The mean

The mean is the sum of all the values in a set of data divided by the number of values in the list. It is what most people understand by the word "average". To find the mean, a simple formula is used.

$$\text{Mean} = \frac{\text{Sum total of values}}{\text{Number of values}}$$

formula to find mean

First, take the list of data and put it in order. Count the number of values in the list. In this example, there are five values.

150, 160, 170, 180, 180

there are five numbers in list

Add all of the values in the list together to find the sum total of the values. In this example the sum total is 840.

add numbers together

$$150 + 160 + 170 + 180 + 180 = 840$$

sum total of values

Divide the sum total of the values, in this case 840, by the number of values, which is 5. The answer, 168, is the mean value of the list.

sum total of values

$$\frac{840}{5} = 168$$

168 is the mean

number of values

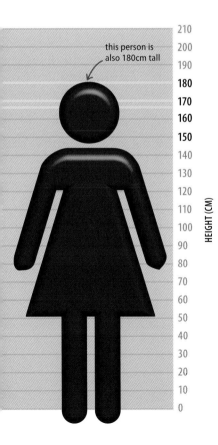

this person is also 180cm tall

HEIGHT (CM)

210 200 190 180 170 160 150 140 130 120 110 100 90 80 70 60 50 40 30 20 10 0

The median

The median is the middle value in a set of data. In a list of five values, it is the third value. In a list of seven values, it would be the fourth value.

median is middle value, in this case the orange figure

Firstly, put the data in ascending order (from lowest to highest)

170, 180, 180, 160, 150

in this list of five values, third value is the median

The median is the middle value in a list with an odd number of values.

150, 160, **170**, 180, 180

Median of an even number of values

In a list with an even number of values, the median is worked out using the two middle values. In a list of six values, these are the third and fourth values.

3rd value   4th value

150, 160, **170**, **180**, 180, 190

middle values

▷ **Calculating the median**
Add the two middle values and divide by two to find the median.

median value

$$\frac{170 + 180}{2} = \frac{350}{2} = 175$$

WORKING WITH FREQUENCY TABLES

Data that deals with averages is often presented in what is known as a frequency table.
Frequency tables show the frequency with which certain values appear in a set of data.

Finding the median using a frequency table

The process for finding the median (middle) value from a frequency table
depends on whether the total frequency is an odd or an even number.

The following marks were scored in a
test and entered in a frequency table:

20, 20, 18, 20, 18, 19, 20, 20, 20

Mark	Frequency
18	2
19	1 (2 + 1 = 3)
20	6 (3 + 6 = 9)
	9

number of
times each ← Frequency
mark appears

median frequency
(entry contains
5th value in list)

total frequency

median
mark

As the total frequency of 9 is odd, to find the median,
first add 1 to it, then divide it by 2, making 5. This means
that the 5th value is the median. Count down the
frequency column adding the values until reaching the
row containing the 5th value. The median mark is 20.

The following marks were scored in a
test and entered in a frequency table:

18, 17, 20 19, 19, 18, 19, 18

Mark	Frequency
17	1
18	3 (1 + 3 = 4)
19	3 (4 + 3 = 7)
20	1 (7 + 1 = 8)
	8

frequency
contains
4th value

frequency
contains
5th value

total frequency

The total frequency of 8 is
even, so there are two middle
values (4th and 5th). Count
down the frequency column
adding values to find them.

▽ **An even total frequency**
If the total frequency is even,
the median is calculated from
the two middle values.

$$\text{Median} = \frac{\text{1st middle value} + \text{2nd middle value}}{2}$$

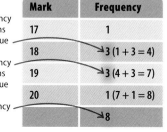

1st middle value

2nd middle value

$$\frac{18 + 19}{2} = \mathbf{18.5}$$

median

The two middle values (4th and 5th)
represent the marks 18 and 19 respectively.
The median is the mean of these two marks,
so add them together and divide by 2. The
median mark is 18.5.

Finding the mean from a frequency table

To find the mean from a frequency table, calculate the total of all the data as well
as the total frequency. Here, the following marks were scored in a test and entered into a table:

16, 18, 20, 19, 17, 19, 18, 17, 18, 19, 16, 19

Mark	Frequency
16	2
17	2
18	3
19	4
20	1

range of values   frequency shows number of
                     times each mark was scored

Mark	Frequency	Total marks (mark × frequency)
16	2	16×2=32
17	2	17×2=34
18	3	18×3=54
19	4	19×4=76
20	1	20×1=20
	12	216

add frequencies together
to get total frequency

total marks

$$\text{Mean} = \frac{\text{Sum of values}}{\text{Number of values}}$$

total marks

total frequency

total marks

$$216 \div 12 = \mathbf{18}$$

total frequency   mean mark

Input the given data into a
frequency table.

Find the total marks scored by multiplying
each mark by its frequency. The total sum of
each part of the data is the sum of values.

To find the mean, divide the sum of values, in
this example, the total marks, by the number of
values, which is the total frequency.

Finding the mean of grouped data

Grouped data is data that has been collected into groups of values, as opposed to specific or individual values. If a frequency table shows grouped data, there is not enough information to calculate the sum of values, so only an estimated value for the mean can be found.

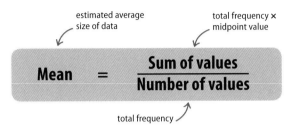

estimated average size of data

total frequency × midpoint value

$$\text{Mean} = \frac{\text{Sum of values}}{\text{Number of values}}$$

total frequency

In grouped data the sum of the values must be found by finding the midpoint of each group and multiplying it by the frequency. Then add each of the results for each group together to find the total frequency × midpoint value. This is divided by the total number of values to find the mean. The example below shows a group of marks scored in a test.

Weighted mean

If some individual values within grouped data contribute more to the mean than other individual values in the group, a "weighted" mean results.

Students in group	15	20	22
Mean exam mark	18	17	13

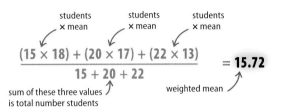

students × mean   students × mean   students × mean

$$\frac{(15 \times 18) + (20 \times 17) + (22 \times 13)}{15 + 20 + 22} = \mathbf{15.72}$$

sum of these three values is total number students

weighted mean

△ **Finding the weighted mean**
Multiply the number of students in each group by the mean mark and add the results. Divide by the total students to give the weighted mean.

Mark	Frequency
under 50	2
50–59	1
60–69	8
70–79	5
80–89	3
90–99	1

Mark	Frequency	Midpoint	Frequency × midpoint
under 50	2	25	2 × 25 = 50
50–59	1	54.5	1 × 54.5 = 54.5
60–69	8	64.5	8 × 64.5 = 516
70–79	5	74.5	5 × 74.5 = 372.5
80–89	3	84.5	3 × 84.5 = 253.5
90–99	1	94.5	1 × 94.5 = 94.5
	20		**1,341**

total frequency

total frequency × midpoint

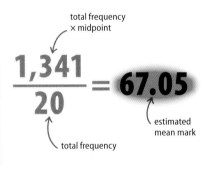

total frequency × midpoint

$$\frac{1{,}341}{20} = \mathbf{67.05}$$

estimated mean mark

total frequency

To find the midpoint of a set of data, add the upper and lower values and divide the answer by 2. For example, the midpoint in the 90–99 mark group is 94.5.

Multiply the midpoint by the frequency for each group and enter this in a new column. Add the results to find the total frequency multiplied by the midpoint.

Dividing the total frequency × midpoint by the total frequency gives the estimated mean mark. It is an estimated value as the exact marks scored are not known – only a range has been given in each group.

The modal class

In a frequency table with grouped data, it is not possible to find the mode (the value that occurs most often in a group). But it is easy to see the group with the highest frequency in it. This group is known as the modal class.

▷ **More than one modal class**
When the highest frequency in the table is in more than one group there is more than one modal class.

Mark	0–25	26–50	51–75	76–100
Frequency	2	6	8	8

modal class

Moving averages

MOVING AVERAGES SHOW GENERAL TRENDS IN DATA
OVER A CERTAIN PERIOD OF TIME.

What is a moving average?

When data is collected over a period of time, the values sometimes
change, or fluctuate, noticeably. Moving averages, or averages over
specific periods of time, smooth out the highs and lows of fluctuating
data and instead show its general trend.

Showing moving averages on a line graph

Taking data from a table, a line graph of individual values over time
can be plotted. The moving averages can also be calculated from the
table data, and a line of moving averages plotted on the same graph.

The table below shows sales of ice cream over a two-year
period, with each year divided into four quarters. The figures for
each quarter show how many thousands of ice creams were sold.

Quarter	YEAR ONE				YEAR TWO			
	1st	2nd	3rd	4th	5th	6th	7th	8th
Sales (in thousands)	1.25	3.75	4.25	2.5	1.5	4.75	5.0	2.75

△ **Table of data**
These figures can be presented as a
line graph, with sales shown on the y
axis and time (measured in quarters
of a year) shown on the x axis.

▷ **Sales graph**
The sales graph shows quarterly
highs and lows (pink line), while a
moving average (green line) shows
the trend over the two-year period.

REAL WORLD

Seasonality

Seasonality is the name given to
regular changes in a data series
that follow a seasonal pattern.
These seasonal fluctuations may
be caused by the weather, or by
annual holiday periods such as
Christmas or Easter. For example,
retail sales experience a predictable
peak around the Christmas period
and low during the summer
holiday period.

▷ **Ice cream sales**
Sales of ice cream tend
to follow a predictable
seasonal pattern.

Calculating moving averages

From the figures in the table, an average for
each period of four quarters can be calculated
and a moving average on the graph plotted.

Average for quarters 1–4
Calculate the mean of the four figures
for year one. Mark the answer on the
graph at the midpoint of the quarters.

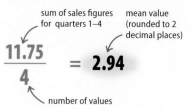

$$1.25 + 3.75 + 4.25 + 2.5 = 11.75$$

sum of sales figures
for quarters 1–4

mean value
(rounded to 2
decimal places)

$$\frac{11.75}{4} = 2.94$$

number of values

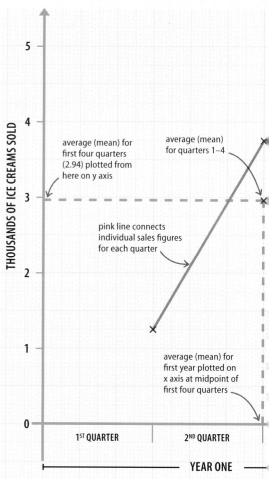

average (mean) for
first four quarters
(2.94) plotted from
here on y axis

average (mean)
for quarters 1–4

pink line connects
individual sales figures
for each quarter

average (mean) for
first year plotted on
x axis at midpoint of
first four quarters

THOUSANDS OF ICE CREAMS SOLD

1ST QUARTER   2ND QUARTER

YEAR ONE

$$\frac{\text{Average}}{\text{(mean)}} = \frac{\text{Sum total of values}}{\text{Number of values}}$$

◁ **Calculating the mean**
Use this formula to find the average (or mean) for each period of four quarters.

Average for quarters 2–5
Calculate the mean of the figures for quarters 2–5 and mark it at the quarters' midpoint.

$$3.75 + 4.25 + 2.5 + 1.5 = 12$$

sum of sales figures for quarters 2–5

mean value

$$\frac{12}{4} = 3$$

number of values

Average for quarters 3–6
Calculate the mean of the figures for quarters 3–6 and mark it at the quarters' midpoint.

$$4.25 + 2.5 + 1.5 + 4.75 = 13$$

sum of sales figures for quarters 3–6

mean value

$$\frac{13}{4} = 3.25$$

number of values

Average for quarters 4–7
Calculate the mean of the figures for quarters 4–7 and mark it at the quarters' midpoint.

$$2.5 + 1.5 + 4.75 + 5 = 13.75$$

sum of sales figures for quarters 4–7

$$\frac{13.75}{4} = 3.44$$

number of values

mean value (rounded to 2 decimal places)

Average for quarters 5–8
Find the mean for quarters 5–8, mark it on the graph, and join all of the marks.

$$1.5 + 4.75 + 5 + 2.75 = 14$$

sum of sales figures for quarters 5–8

mean value

$$\frac{14}{4} = 3.5$$

number of values

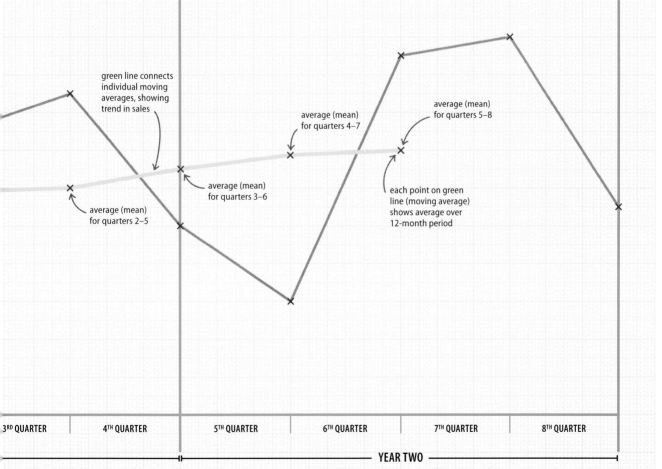

green line connects individual moving averages, showing trend in sales

average (mean) for quarters 4–7

average (mean) for quarters 5–8

average (mean) for quarters 3–6

average (mean) for quarters 2–5

each point on green line (moving average) shows average over 12-month period

3ᴿᴰ QUARTER   4ᵀᴴ QUARTER   5ᵀᴴ QUARTER   6ᵀᴴ QUARTER   7ᵀᴴ QUARTER   8ᵀᴴ QUARTER

YEAR TWO

Measuring spread

MEASURES OF SPREAD SHOW THE RANGE OF DATA, AND ALSO GIVE
MORE INFORMATION ABOUT THE DATA THAN AVERAGES ALONE.

SEE ALSO

❮ **196–197** Collecting and
organizing data

Histograms **216–217** ❯

Diagrams showing the measure of spread give the highest and lowest figures
– the range – of the data and give information about how it is distributed.

Range and distribution

From tables or lists of data, diagrams can be created
that show the ranges of different sets of data. This
shows the distribution of the data, whether it is
spread over a wide or narrow range.

Subject	Ed's results	Bella's results
Maths	47	64
English	95	68
French	10	72
Geography	65	61
History	90	70
Physics	60	65
Chemistry	81	60
Biology	77	65

This table shows the marks of two students. Although
their average (see pp.206–207) marks are the same
(65.625), the ranges of their marks are very different.

REAL WORLD

Broadband bandwidth

Internet service providers
often give a maximum
speed for their broadband
connections, for example
20Mb per second. However,
this information can be
misleading. An average
speed gives a better idea
of what to expect, but the
range and distribution
of the data is the
information really
needed to get the
full picture.

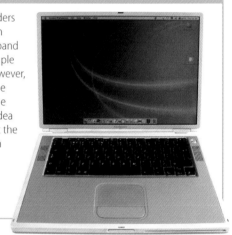

lowest mark

highest mark

Ed: **10**, 47, 60, 65, 77, 81, 90, **95**

Bella: **60**, 61, 64, 65, 65, 68, 70, **72**

◁ **Finding the range**
To calculate the range of each student's
marks, subtract the lowest figure from
the highest in each set. Ed's lowest mark
is 10, and highest 95, so his range is 85.
Bella's lowest mark is 60, and highest 72,
giving a range of 12.

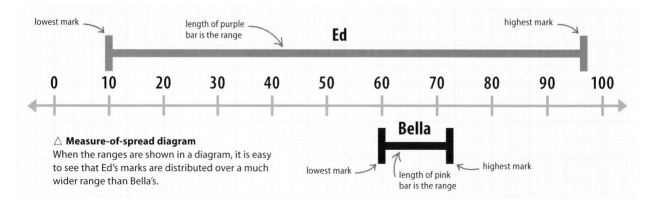

lowest mark

length of purple
bar is the range

Ed

highest mark

0   10   20   30   40   50   60   70   80   90   100

Bella

lowest mark

length of pink
bar is the range

highest mark

△ **Measure-of-spread diagram**
When the ranges are shown in a diagram, it is easy
to see that Ed's marks are distributed over a much
wider range than Bella's.

Stem-and-leaf diagrams

Another way of showing data is in stem-and-leaf diagrams. These give a clearer picture of the way the data is distributed within the range than a simple measure-of-spread diagram.

This is how the data appears before it has been organized.

34, 48, 7, 15, 27, 18, 21, 14, 24, 57, 25, 12, 30, 37, 42, 35, 3, 43, 22, 34, 5, 43, 45, 22, 49, 50, 34, 12, 33, 39, 55

Sort the list of data into numerical order, with the smallest number first. Add a zero in front of any number smaller than 10.

03, 05, 07, 12, 12, 14, 15, 18, 21, 22, 22, 24, 25, 27, 30, 33, 34, 34, 34, 35, 37, 39, 42, 43, 43, 45, 48, 49, 50, 55, 57

To draw a stem-and-leaf diagram, draw a cross with more space to the right of it than the left. Write the data into the cross, with the tens in the "stem" column to the left of the cross, and the units for each number as the "leaves" on the right hand side. Once each value of tens has been entered into the stem, do not repeat it, but continue to repeat the values entered into the leaves.

this is the stem. 1 stands for 10, 2 for 20, and so on

this is the leaf, which is joined to the stem to form a complete number

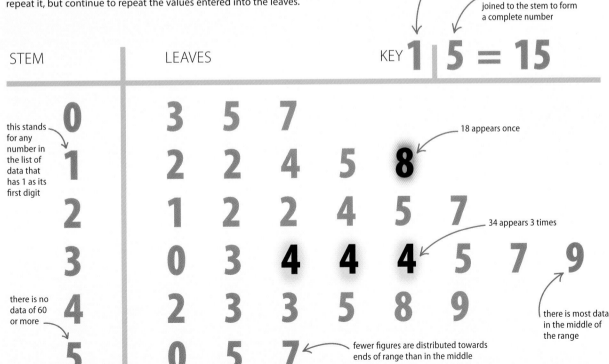

STEM

LEAVES

KEY 1 | 5 = 15

this stands for any number in the list of data that has 1 as its first digit

0   3   5   7

1   2   2   4   5   **8**   — 18 appears once

2   1   2   2   4   5   7   — 34 appears 3 times

3   0   3   **4**   **4**   **4**   5   7   9

there is no data of 60 or more

4   2   3   3   5   8   9

there is most data in the middle of the range

5   0   5   7   — fewer figures are distributed towards ends of range than in the middle

QUARTILES

Quartiles are dividing points in the range of a set of data that give a clear picture of distribution. The median marks the centre point, the upper quartile marks the midpoint between the median and the top of the distribution, and the lower quartile the midpoint between the median and the bottom. Estimates of quartiles can be found from a graph, or calculated precisely using formulas.

Estimating quartiles

Quartiles can be estimated by reading values from a cumulative frequency graph (see p.205).

Make a table with the data given for range and frequency, and add up the cumulative frequency. Use this data to make a cumulative frequency graph, with cumulative frequency on the y axis, and range on the x axis.

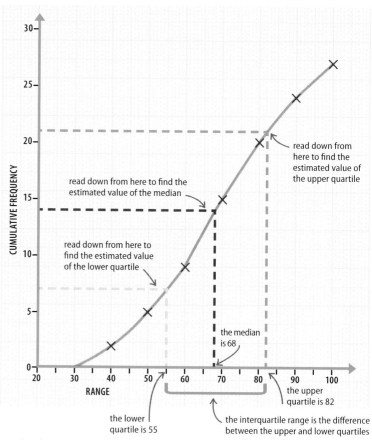

read down from here to find the estimated value of the median

read down from here to find the estimated value of the upper quartile

read down from here to find the estimated value of the lower quartile

the median is 68

the lower quartile is 55

the upper quartile is 82

the interquartile range is the difference between the upper and lower quartiles

Range	Frequency	Cumulative frequency
30–39	2	2
40–49	3	5 (2+3)
50–59	4	9 (2+3+4)
60–69	6	15 (2+3+4+6)
70–79	5	20 (2+3+4+6+5)
80–89	4	24 (2+3+4+6+5+4)
>90	3	27 (2+3+4+6+5+4+3)

this sign means greater than

add each number to those before it to find cumulative frequency

▶ **Divide the total cumulative frequency** by 4 (this will be the cumulative frequency of the last entry in the table), and use the result to divide the y axis into 4 parts.

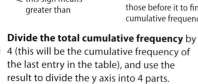

total cumulative frequency

divide the y axis into sections of this length

$$\frac{27}{4} = 6.75$$

▶ **Read across from the marks** and down to the x axis to find estimated values for the quartiles. These are only approximate values.

Calculating quartiles

Exact values of quartiles can be found from a list of data. These formulas give the position of the quartiles and median in a list of data in ascending order, using the total number of data items in the list, n.

n is the total number of values in the list

$$\frac{(n + 1)}{4}$$

△ **Lower quartile**
This shows the position of the lower quartile in a list of data.

$$\frac{(n + 1)}{2}$$

△ **Median**
This shows the position of the median in a list of data.

$$\frac{3 (n + 1)}{4}$$

△ **Upper quartile**
This shows the position of the upper quartile in a list of data.

How to calculate quartiles

To find the values of the quartiles in a list of data, first arrange the
list of numbers in ascending order from lowest to highest.

37,38,45,47,48,51,54,54,58,60,62,63,63,65,69,71,74,75,78,78,80,84,86,89,92,94,96

▶ **Using the formulas,** calculate where to find the
quartiles and the median in this list. The answers
give the position of each value in the list.

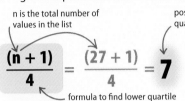

n is the total number of
values in the list

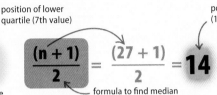

position of lower
quartile (7th value)

position of median
(14th value)

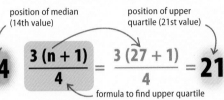

position of upper
quartile (21st value)

$$\frac{(n+1)}{4} = \frac{(27+1)}{4} = \mathbf{7}$$

formula to find lower quartile

$$\frac{(n+1)}{2} = \frac{(27+1)}{2} = \mathbf{14}$$

formula to find median

$$\frac{3\,(n+1)}{4} = \frac{3\,(27+1)}{4} = \mathbf{21}$$

formula to find upper quartile

△ **Lower quartile**
This calculation gives the answer 7,
so the lower quartile is the 7th value
in the list.

△ **Median**
The answer to this calculation is
14, so the median is the 14th value
in the list.

△ **Upper quartile**
The answer to this calculation is
21, so the upper quartile is the
21st value in the list.

▶ **To find the values** of the quartiles and the
median, count along the list to the positions
that have just been calculated.

lower quartile   median   upper quartile

| 1 | 2 | 3 | 4 | 5 | 6 | 7 | 8 | 9 | 10 | 11 | 12 | 13 | 14 | 15 | 16 | 17 | 18 | 19 | 20 | 21 | 22 | 23 | 24 | 25 | 26 | 27 |

37,38,45,47,48,51,**54**,54,58,60,62,63,63,**65**,69,71,74,75,78,78,**80**,84,86,89,92,94,96

LOOKING CLOSER

Box-and-whisker diagram

Box-and-whisker diagrams are a way of showing the spread and
distribution of a range of data in an graphic way. The range is
plotted on a number line, with the interquartile range between
the upper and lower quartiles shown as a box.

▽ **Using the diagram**
This box-and-whisker diagram shows a
range with a lower limit of 1 and an
upper limit of 9. The median is 4, the
lower quartile 3, and the upper quartile 6.

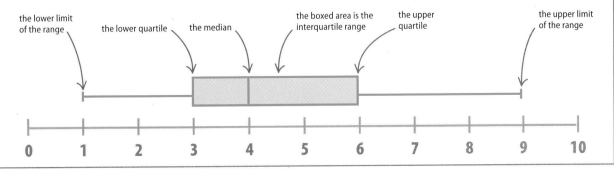

the lower limit
of the range

the lower quartile

the median

the boxed area is the
interquartile range

the upper
quartile

the upper limit
of the range

Histograms

A HISTOGRAM IS A TYPE OF BAR CHART. IN A HISTOGRAM, THE AREA OF THE BARS, RATHER THAN THEIR LENGTH, REPRESENTS THE SIZE OF THE DATA.

SEE ALSO

❰ **196–197** Collecting and organizing data

❰ **198–201** Bar charts

❰ **212–215** Measuring spread

What is a histogram?

A histogram is a diagram made up of blocks on a graph. Histograms are useful for showing data when it is grouped into groups of different sizes. This example looks at the number of downloads of a music file in a month (frequency) by different age groups. Each age group (class) is a different size because each covers a different age range. The width of each block represents the age range, known as class width. The height of each block represents frequency density, which is calculated by dividing the number of downloads (frequency) in each age group (class) by the class width (age range).

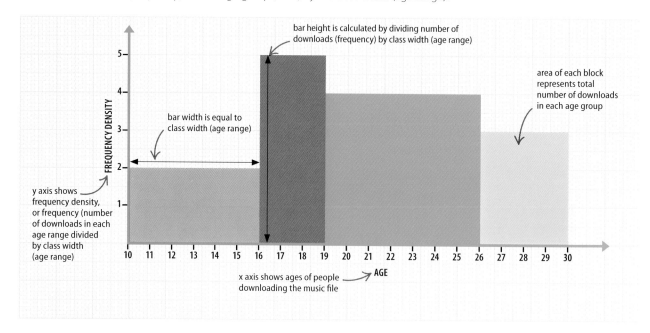

bar height is calculated by dividing number of downloads (frequency) by class width (age range)

area of each block represents total number of downloads in each age group

bar width is equal to class width (age range)

y axis shows frequency density, or frequency (number of downloads in each age range divided by class width (age range)

x axis shows ages of people downloading the music file

LOOKING CLOSER

Histograms and bar charts

Bar charts look like histograms, but show data in a different way. In a bar chart, the bars are all the same width. The height of each bar represents the total (frequency) for each group, while in a histogram, totals are represented by the area of the blocks.

▷ **Bar chart**

This bar chart shows the same data as shown above. Although class widths are different, the widths of the bars are all the same.

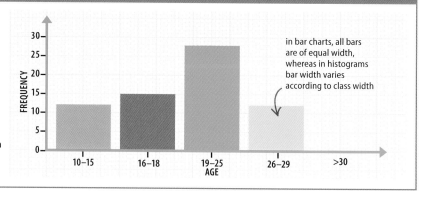

in bar charts, all bars are of equal width, whereas in histograms bar width varies according to class width

How to draw a histogram

To draw a histogram, begin by making a frequency table for the data. Next, using the class boundaries, find the width of each class of data. Then calculate frequency density for each by dividing frequency by class width.

upper class boundary for any group is lower boundary of next group

class boundaries for this data are 10, 16, 19, 26, and 30

find class width by subtracting lower class boundary from the upper class boundary, for example 16 – 10 = 6

number of downloads per month

divide frequency by class width to find frequency density

Age (year)	Frequency (downloads in a month)
10–15	12
16–18	15
19–25	28
26–29	12
>30	0

Age	Class width	Frequency	Frequency density
10–15	6	12	2
16–18	3	15	5
19–25	7	28	4
26–29	4	12	3
>30	–	0	–

there is no data to enter for this group

The information needed to draw a histogram is the range of each class of data and frequency data. From this information, the class width and frequency density can be calculated.

To find class width, begin by finding the class boundaries of each group of data. These are the two numbers that all the values in a group fall in between – for example, for the 10–15 group they are 10 and 16. Next, find class width by subtracting the lower boundary from the upper for each group.

To find frequency density, divide the frequency by the class width of each group. Frequency density shows the frequency of each group in proportion to its class width.

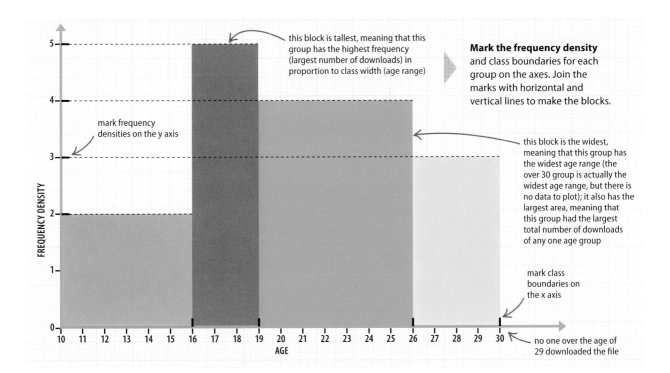

this block is tallest, meaning that this group has the highest frequency (largest number of downloads) in proportion to class width (age range)

mark frequency densities on the y axis

Mark the frequency density and class boundaries for each group on the axes. Join the marks with horizontal and vertical lines to make the blocks.

this block is the widest, meaning that this group has the widest age range (the over 30 group is actually the widest age range, but there is no data to plot); it also has the largest area, meaning that this group had the largest total number of downloads of any one age group

mark class boundaries on the x axis

no one over the age of 29 downloaded the file

FREQUENCY DENSITY

AGE

Scatter diagrams

SCATTER DIAGRAMS PRESENT INFORMATION FROM TWO SETS OF
DATA AND REVEAL THE RELATIONSHIP BETWEEN THEM.

SEE ALSO

❮ **196–197** Collecting and organizing data
❮ **204–205** Line graphs

What is a scatter diagram?

A scatter diagram is a graph made from two sets of data. Each set of data is
measured on an axis of the graph. The data always appears in pairs – one value
will need to be read up from the x axis, the other read across from the y axis. A
point is marked where each pair meet. The pattern made by the points shows
whether there is any connection, or correlation, between the two sets of data.

▽ **Table of data**
This table shows two sets of
data – the height and weight of
13 people. With each person's
height their corresponding
weight measurement is given.

Height (cm)	173	171	189	167	183	181	179	160	177	180	188	186	176
Weight (kg)	69	68	90	65	77	76	74	55	70	75	86	81	68

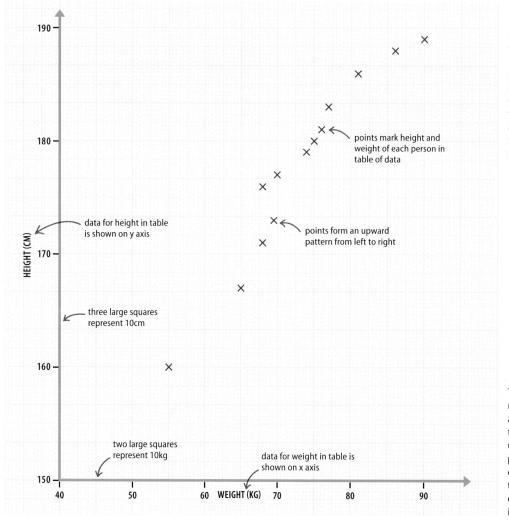

◁ **Plotting the points**
Draw a vertical axis (y)
and a horizontal axis (x)
on graph paper. Mark out
measurements for each
set of data in the table
along the axes. Read each
corresponding height and
weight in from its axis
and mark where they
meet. Do not join the
points marked.

points mark height and
weight of each person in
table of data

data for height in table
is shown on y axis

points form an upward
pattern from left to right

three large squares
represent 10cm

two large squares
represent 10kg

data for weight in table is
shown on x axis

◁ **Positive correlation**
The pattern of points
marked between the two
axes shows an upward
trend from left to right. An
upward trend is known as
positive correlation. The
correlation between the
two sets of data in this
example is that as height
increases, so does weight.

Negative and zero correlations

The points in a scatter diagram can form many different patterns, which reveal different types of correlation between the sets of data. This can be positive, negative, or non-existent. The pattern can also reveal how strong or how weak the correlation is between the two sets of data.

Energy used (kwh)	1,000	1,200	1,300	1,400	1,450	1,550	1,650	1,700
Temperature (°C)	55	50	45	40	35	30	25	20

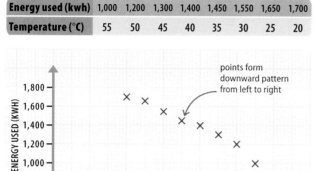

IQ	141	127	117	150	143	111	106	135
Shoe size	8	10	11	6	11	10	9	7

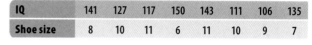
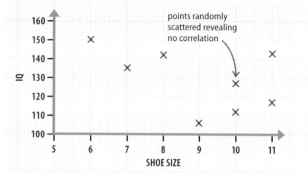

△ **Negative correlation**
In this graph, the points form a downward pattern from left to right. This reveals a connection between the two sets of data – as the temperature increases, energy consumption goes down. This relationship is called negative correlation.

△ **No correlation**
In this graph, the points form no pattern at all – they are widely spaced and do not reveal any trend. This shows there is no connection between a person's shoe size and their IQ, which means there is zero correlation between the two sets of data.

Line of best fit

To make a scatter diagram clearer and easier to read, a straight line can be drawn that follows the general pattern of the points, with an equal number of points on both sides of the line. This line is called the line of best fit.

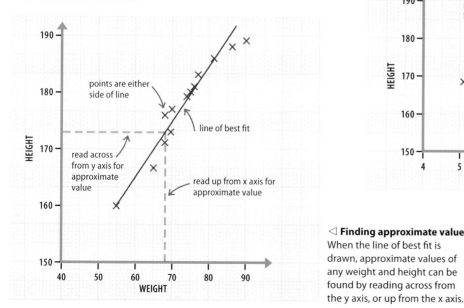

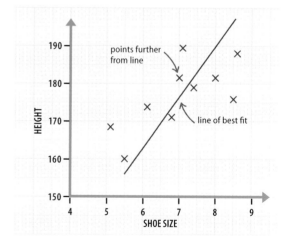

◁ **Finding approximate values**
When the line of best fit is drawn, approximate values of any weight and height can be found by reading across from the y axis, or up from the x axis.

△ **Weak correlation**
Here the points are further away from the line of best fit. This shows that the correlation between height and shoe size is weak. The further the points are from the line, the weaker the correlation.

Probability

What is probability?

PROBABILITY IS THE LIKELIHOOD OF SOMETHING HAPPENING.

Maths can be used to calculate the likelihood or chance that something will happen.

SEE ALSO

❰ **40–47** Fractions

❰ **56–57** Converting fractions, decimals, and percentages

Expectation and reality **224–225 ❱**

Combined probabilities **226–227 ❱**

How is probability shown?

Probabilities are given a value between 0, which is impossible, and 1, which is certain. To calculate these values, fractions are used. Follow the steps to find out how to calculate the probability of an event happening and then how to show it as a fraction.

total of specific events that can happen

$$\frac{1}{8}$$

◁ **Writing a probability**
The top number shows the chances of a specific event, while the bottom number shows the total chances of all of the possible events happening.

total of all possible events that can happen

▷ **Total chances**
Decide what the total number of possible outcomes is. In this example, with 5 sweets to pick 1 sweet from, the total is 5, as any one of 5 sweets may be picked.

there are 5 sweets, 4 are red and 1 is yellow

▷ **Chance of red sweet**
Of the 5 sweets, 4 are red. This means that there are 4 chances out of 5 that the sweet chosen is red. This probability can be written as a fraction ⅘.

total number of red sweets that can be chosen

$$\frac{4}{5}$$

total of 5 sweets to choose from

▷ **Chance of yellow sweet**
As 1 sweet is yellow there is 1 chance in 5 of the sweet picked being yellow. This probability can be written as a fraction ⅕.

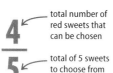

1 yellow sweet can be chosen

$$\frac{1}{5}$$

total of 5 sweets to choose from

△ **Identical snowflakes**
Every snowflake is unique and the chance that there can be two identical snowflakes is 0 on the scale, or impossible.

▷ **A hole in one**
A hole in one during a game of golf is highly unlikely, so it has a probability close to 0 on the scale. However, it can still happen!

0

IMPOSSIBLE

UNLIKELY

▷ **Probability scale**
All probabilities can be shown on a line known as a probability scale. The more likely something is to occur the further to the right, or towards 1, it is placed.

LESS LIKELY

Calculating probabilities

This example shows how to work out the probability of randomly picking a red sweet from a group of 10 sweets. The number of ways this event could happen is put at the top of the fraction and the total number of possible events is put at the bottom.

number of red sweets that can be chosen

$$\frac{3 \text{ red sweets}}{10 \text{ sweets}}$$

total that can be chosen

chance of red sweet being chosen, as fraction

$$\frac{3}{10} \text{ or } 0.3$$

chance of red sweet being chosen, as decimal

△ **Pick a sweet**
There are 10 sweets to choose from. Of these, 3 are coloured red. If one of the sweets is picked, what is the chance of it being red?

△ **Red randomly chosen**
One sweet is chosen at random from the 10 coloured sweets. The sweet chosen is one of the 3 red sweets available.

△ **Write as a fraction**
There are three reds that can be chosen, so 3 is put at the top of the fraction. As there are ten sweets in total, 10 is at the bottom.

△ **What is the chance?**
The probability of a red sweet being picked is 3 out of 10. This can be written as the fraction ³⁄₁₀, or the decimal 0.3.

◁ **Heads or tails**
If a coin is tossed there is a 1 in 2, or even, chance of throwing either a head or a tail. This is shown as 0.5 on the scale, which is the same as half, or 50%.

▷ **Earth turning**
It is a certainty that each day the Earth will continue to turn on its axis, making it a 1 on the scale.

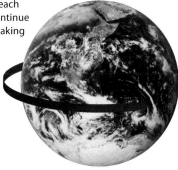

vast majority of people are right-handed

◁ **Being right-handed**
The chances of picking at random a right-handed person are very high – almost 1 on the scale. Most people are right-handed.

 0.5
EVEN CHANCE

LIKELY

1
CERTAIN

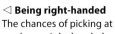 **MORE LIKELY**

Expectation and reality

EXPECTATION IS AN OUTCOME THAT IS ANTICIPATED TO OCCUR; REALITY IS THE OUTCOME THAT ACTUALLY OCCURS.

SEE ALSO

❰ **40–47** Fractions

❰ **222–223** What is probability?

Combined probabilities **226–227** ❱

The difference between what is expected to occur and what actually occurs can often be considerable.

What is expectation?

There is an equal chance of a 6-sided dice landing on any number. It is therefore expected that each of the 6 numbers on it will be rolled once in every 6 throws ($\frac{1}{6}$ of the time). Similarly, if a coin is tossed twice, it is expected that it will land on heads once and tails once. However, this does not always happen in real life.

WHAT ARE THE CHANCES?	
Two random phone numbers ending in same digit	1 chance in 10
Randomly selected person being left-handed	1 chance in 12
Pregnant woman giving birth to twins	1 chance in 33
An adult living to 100	1 chance in 50
A random clover having four leaves	1 chance in 10,000
Being struck by lightning in a year	1 chance in 2.5 million
A specific house being hit by a meteor	1 chance in 182 trillion

chance of rolling each number is 1 in 6

△ **Roll a dice**
Roll a dice 6 times and it seems likely that each of the 6 numbers on the dice will be seen once.

Expectation versus reality

Mathematical probability expects that when a dice is rolled 6 times, the numbers 1, 2, 3, 4, 5, and 6 will appear once each, but it is unlikely this outcome would actually occur. However, over a longer series of events, for example, throwing a dice a thousand times, the total numbers of 1s, 2s, 3s, 4s, 5s, and 6s thrown would be more even.

reasonable to expect 4 in first 6 throws

▷ **Expectation**
Mathematical probability expects that, when a dice is rolled 6 times, a 4 will be thrown once.

unexpected third 5 in 6 throws

unexpected third 6 in 6 throws

▷ **Reality**
Throwing a dice 6 times may create any combination of the numbers on a dice.

Calculating expectation

Expectation can be calculated. This is done by expressing the likelihood of something happening as a fraction, and then multiplying the fraction by the number of times the occurrence has the chance to happen. This example shows how expectation can be calculated in a game where balls are pulled from a bucket, with numbers ending in 0 or 5 winning a prize.

◁ **Numbered balls**
There are 30 balls in the bucket and 5 are removed at random. The balls are then checked for winning numbers – numbers that end in 0 or 5.

6 winning balls

5 10 15 20 25 30

30 numbered balls to pick from

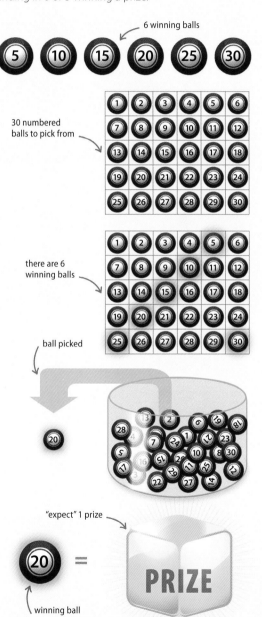

there are 6 winning balls

ball picked

20

"expect" 1 prize

(20) = **PRIZE**

winning ball

There are 6 winning balls that can be picked out of the total of 30 balls.

number of winning balls in game

6

The total number of balls that can be picked in the game is 30.

total number of balls in game

30

both parts of the fraction can divide by 6, so it can be cancelled

chances of winning ball being picked
$6 \div 6 = 1$

The probability of a winning ball being picked is 6 (balls) out of 30 (balls). This can be written as the fraction $6/30$, and is then cancelled to $1/5$. The chance of picking a winning ball is 1 in 5, so the chance of winning a prize is 1 in 5.

$$\frac{6}{30} = \frac{1}{5}$$

$30 \div 6 = 5$

1 prize "probably" won

It is expected that a prize will be won exactly 1 out of 5 times. The probability of winning a prize is therefore $1/5$ of 5, which is 1.

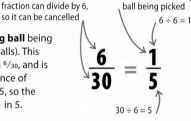

$$\frac{1}{5} \times 5 = 1$$

probability of picking a winning ball is 1 in 5

opportunities to pick a ball

Expectation suggests that 1 prize will be won if 5 balls are picked. However, in reality no prize or even 5 prizes could be won.

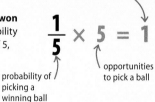

1 prize won?

Combined probabilities

THE PROBABILITY OF ONE OUTCOME FROM TWO OR MORE EVENTS
HAPPENING AT THE SAME TIME, OR ONE AFTER THE OTHER.

SEE ALSO

❮ **222–223** What is
probability?
❮ **224–225** Expectation
and reality

Calculating the chance of one outcome from two things happening at the
same time is not as complex as it might appear.

What are combined probabilities?

To find out the probability of one possible outcome
happening from more than one event, all of the possible
outcomes need to be worked out first. For example, if a
coin is tossed and a dice is rolled at the same time, what
is the probability of the coin landing on tails and the
dice rolling a 4?

coin has 2 sides   dice has 6 sides

COIN   DICE

Coin and dice
A coin has 2 sides
(heads and tails) while
a dice has 6 sides –
numbers 1 through to 6,
represented by numbers
of dots on each side.

▷ **Tossing a coin**
As there are 2 sides to a
coin, and each is equally
likely to show if the coin is
tossed. This means that the
chance of the coin landing
on tails is exactly 1 in 2,
shown as the fraction ½.

chance of heads is
1 in 2

chance of tails is
1 in 2

HEADS   TAILS

$\dfrac{1}{2}$

represents
chance of
single event, for
example chance
of coin landing
on tails

represents total
possible
outcomes if
coin is tossed

▷ **Rolling a dice**
As there are 6 sides to a
dice, and each side is
equally likely to show
when the dice is rolled,
the chance of rolling a 4
is exactly 1 in 6, shown
as the fraction ⅙.

chance of rolling
1 is 1 in 6

chance of rolling
2 is 1 in 6

chance of rolling
3 is 1 in 6

chance of
rolling 4 is
1 in 6

chance of
rolling 5 is
1 in 6

chance of rolling
6 is 1 in 6

$\dfrac{1}{6}$

represents
chance of
single event,
for example
chance of
rolling a 4

represents total
possible
outcomes if dice
is thrown

▷ **Both events**
To find out the chances
of both a coin landing
on tails and a dice
simultaneously rolling a
4, multiply the individual
probabilities together.
The answer shows that
there is a ½ chance of
this outcome.

coin lands on tails

multiply the 2
probabilities together

chance of dice rolling
a 4 is 1 in 6

chance of specific
outcome

chance of coin
landing on tails
and rolling
a 4 is 1 in 12

total possible
outcomes

TAILS

chance of coin landing
on tails is 1 in 2

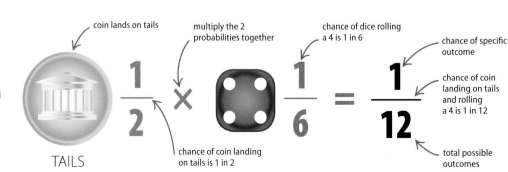

$$\frac{1}{2} \times \frac{1}{6} = \frac{1}{12}$$

Working out possible outcomes

A table can be used to work out all the possible outcomes of two combined events. For example, if two dice are rolled their scores will have a combined total of between 2 and 12. There are 36 possible outcomes, which are shown in the table below. Read down from each red dice and across from each blue dice for each of their combined results.

blue dice throws

red dice throws

Red / Blue	⚀	⚁	⚂	⚃	⚄	⚅
⚀	2	3	4	5	6	7
⚁	3	4	5	6	7	8
⚂	4	5	6	7	8	9
⚃	5	6	7	8	9	10
⚄	6	7	8	9	10	11
⚅	7	8	9	10	11	12

6 ways out of 36 to throw 7, for example blue dice rolling 1 and red dice rolling 6

5 ways out of 36 to throw 8, for example blue dice rolling 2 and red dice rolling 6

4 ways out of 36 to throw 9, for example blue dice rolling 3 and red dice rolling 6

3 ways out of 36 to throw 10, for example blue dice rolling 4 and red dice rolling 6

2 ways out of 36 to throw 11, for example blue dice rolling 5 and red dice rolling 6

1 way out of 36 to throw 12

KEY

 Least likely
The least likely outcome of throwing 2 dice is either 2 (each dice is 1) or 12 (each is 6). There is a $\frac{1}{36}$ chance of either result.

 Most likely
The most likely outcome of throwing 2 dice is a 7. With 6 ways to throw a 7, there is a $\frac{6}{36}$, or $\frac{1}{6}$, chance of this result.

Dependent events

SEE ALSO

❮ **224–225** Expectation and reality

THE CHANCES OF SOMETHING HAPPENING CAN CHANGE ACCORDING TO THE EVENTS THAT PRECEDED IT. THIS IS A DEPENDENT EVENT.

Dependent events

In this example, the probability of picking any one of four green cards from a pack of 40 is 4 out of 40 (4/40). It is an independent event. However, the probability of the second card picked being green depends on the colour of the card picked first. This is known as a dependent event.

▷ **Colour-coded**
This pack of cards contains 10 groups, each with its own colour. There are 4 cards in each group.

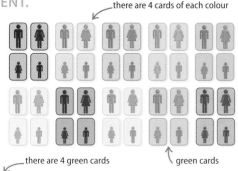

there are 4 cards of each colour

there are 4 green cards

green cards

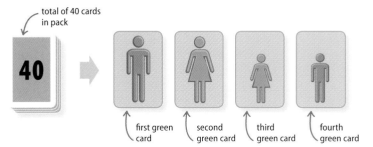

total of 40 cards in pack

first green card

second green card

third green card

fourth green card

$$\frac{4}{40}$$

◁ **What are the chances?**
The chances of the first card picked being green is 4 in 40 (4/40). This is independent of other events because it is the first event.

there are 40 cards in total

Dependent events and decreasing probability

If the first card chosen from a pack of 40 is one of the 4 green cards, then the chances that the next card is green are reduced to 3 in 39 (3/39). This example shows how the chances of a green card being picked next gradually shrink to zero.

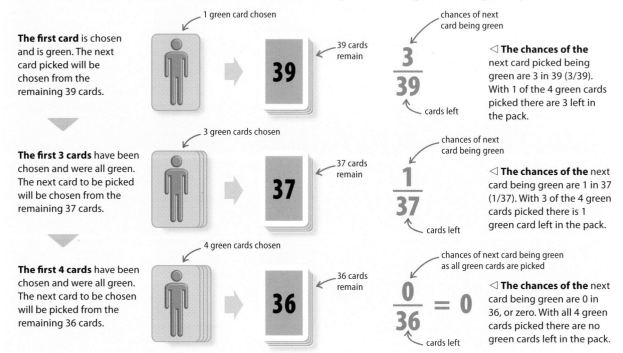

The first card is chosen and is green. The next card picked will be chosen from the remaining 39 cards.

1 green card chosen

39 cards remain

chances of next card being green

$$\frac{3}{39}$$

cards left

◁ **The chances of the** next card picked being green are 3 in 39 (3/39). With 1 of the 4 green cards picked there are 3 left in the pack.

The first 3 cards have been chosen and were all green. The next card to be picked will be chosen from the remaining 37 cards.

3 green cards chosen

37 cards remain

chances of next card being green

$$\frac{1}{37}$$

cards left

◁ **The chances of the** next card being green are 1 in 37 (1/37). With 3 of the 4 green cards picked there is 1 green card left in the pack.

The first 4 cards have been chosen and were all green. The next card to be chosen will be picked from the remaining 36 cards.

4 green cards chosen

36 cards remain

chances of next card being green as all green cards are picked

$$\frac{0}{36} = 0$$

cards left

◁ **The chances of the** next card being green are 0 in 36, or zero. With all 4 green cards picked there are no green cards left in the pack.

Dependent events and increasing probability

If the first card chosen from a pack of 40 is not one of the 4 pink cards, then the probability of the next card being pink grow to 4 out of the remaining 39 cards (4/39). In this example, the probability of a pink card being the next to be picked grows to a certainty with each non-pink card picked.

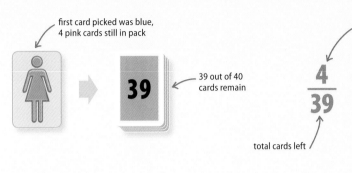

first card picked was blue, 4 pink cards still in pack

The first card has been chosen and is not pink. The next card to be picked will be chosen from the remaining 39 cards.

39 out of 40 cards remain

$\frac{4}{39}$

chances of next card being pink

total cards left

◁ **The chances of the** next card being pink are 4 out of 39 (4/39). This is because none of the 4 pink cards were picked so there are 4 still left in the pack.

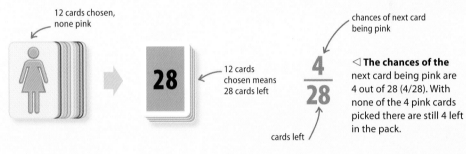

12 cards chosen, none pink

The first 12 cards have been chosen, none of which were pink. The next card to be picked will be chosen from the remaining 28 cards.

12 cards chosen means 28 cards left

$\frac{4}{28}$

chances of next card being pink

cards left

◁ **The chances of the** next card being pink are 4 out of 28 (4/28). With none of the 4 pink cards picked there are still 4 left in the pack.

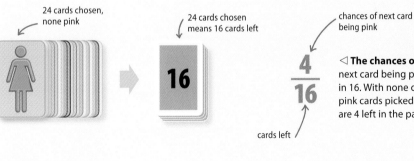

24 cards chosen, none pink

24 cards have been chosen and none of which were pink. The next card to be picked will be chosen from the remaining 16 cards.

24 cards chosen means 16 cards left

$\frac{4}{16}$

chances of next card being pink

cards left

◁ **The chances of the** next card being pink are 4 in 16. With none of the 4 pink cards picked there are 4 left in the pack.

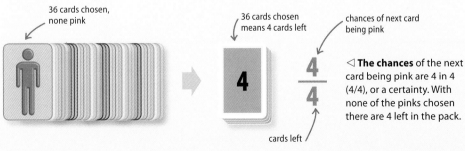

36 cards chosen, none pink

The first 36 cards have been chosen. None of them were pink. The next card to be picked will be chosen from the remaining 4 cards.

36 cards chosen means 4 cards left

$\frac{4}{4}$

chances of next card being pink

cards left

◁ **The chances** of the next card being pink are 4 in 4 (4/4), or a certainty. With none of the pinks chosen there are 4 left in the pack.

Tree diagrams

TREE DIAGRAMS CAN BE CONSTRUCTED TO HELP CALCULATE THE
PROBABILITY OF MULTIPLE EVENTS OCCURRING.

SEE ALSO

‹ 222–223 What is probability?

‹ 226–227 Combined probabilities

‹ 228–229 Dependent events

A range of probable outcomes of future events can be shown using arrows, or the "branches" of a "tree", flowing from left to right.

Building a tree diagram

The first stage of building a tree diagram is to draw an arrow from the start position to each of the possible outcomes. In this example, the start is a mobile phone, and the outcomes are 5 messages sent to 2 other phones, with each of these other phones at the end of 1 of 2 arrows. As no event came before, they are single events.

▷ **Single events**
Of 5 messages, 2 are sent to the first phone, shown by the fraction ⅖, and 3 out of 5 are sent to the second phone, shown by the fraction ⅗.

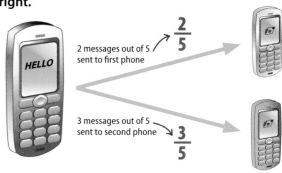

2 messages out of 5 sent to first phone $\frac{2}{5}$

3 messages out of 5 sent to second phone $\frac{3}{5}$

Tree diagrams showing multiple events

To draw a tree diagram that shows multiple events, begin with a start position, with arrows leading to the right to each of the possible outcomes. This is stage 1. Each of the outcomes of stage 1 then becomes a new start position, with further arrows each leading to a new stage of possible outcomes. This is stage 2. More stages can then follow on from the outcomes of previous stages. As one stage of events comes before another, these are multiple events.

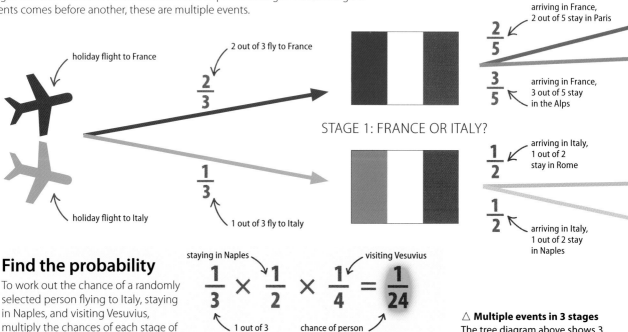

holiday flight to France

2 out of 3 fly to France

$\frac{2}{3}$

STAGE 1: FRANCE OR ITALY?

arriving in France, 2 out of 5 stay in Paris $\frac{2}{5}$

arriving in France, 3 out of 5 stay in the Alps $\frac{3}{5}$

arriving in Italy, 1 out of 2 stay in Rome $\frac{1}{2}$

arriving in Italy, 1 out of 2 stay in Naples $\frac{1}{2}$

$\frac{1}{3}$

holiday flight to Italy

1 out of 3 fly to Italy

Find the probability

To work out the chance of a randomly selected person flying to Italy, staying in Naples, and visiting Vesuvius, multiply the chances of each stage of this journey together for the answer.

staying in Naples

visiting Vesuvius

$$\frac{1}{3} \times \frac{1}{2} \times \frac{1}{4} = \frac{1}{24}$$

1 out of 3 fly to Italy

chance of person visiting Italy, then Naples, then Vesuvius

△ **Multiple events in 3 stages**
The tree diagram above shows 3 stages of a holiday. In stage 1, people fly to France or Italy.

When multiple events are dependent

Tree diagrams show how the chances of one event can depend on the previous event. In this example, each event is someone picking a fruit from a bag and not replacing it.

△ **Dependent events**
The first person picks from a bag of 10 fruits (3 oranges, 7 apples). The next picks from 9 fruits, when the chances of what is picked are out of 9.

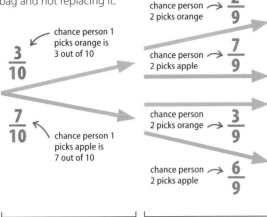

chance person 1 picks orange is 3 out of 10 — $\frac{3}{10}$

chance person 1 picks apple is 7 out of 10 — $\frac{7}{10}$

chance person 2 picks orange → $\frac{2}{9}$

chance person 2 picks apple → $\frac{7}{9}$

chance person 2 picks orange → $\frac{3}{9}$

chance person 2 picks apple → $\frac{6}{9}$

person 1 chooses from 10 fruits | person 2 chooses from 9 fruits

Find the probability

What are the chances that the first and second person will each choose an orange? Multiply the chances of both events together.

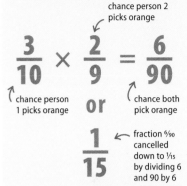

chance person 2 picks orange

$$\frac{3}{10} \times \frac{2}{9} = \frac{6}{90}$$

chance person 1 picks orange

or

chance both pick orange

$$\frac{1}{15}$$

fraction ⁶⁄₉₀ cancelled down to ¹⁄₁₅ by dividing 6 and 90 by 6

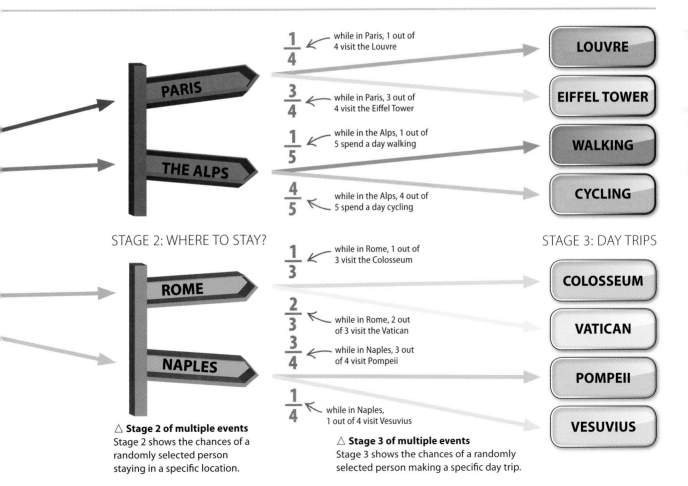

$\frac{1}{4}$ ← while in Paris, 1 out of 4 visit the Louvre → **LOUVRE**

$\frac{3}{4}$ ← while in Paris, 3 out of 4 visit the Eiffel Tower → **EIFFEL TOWER**

$\frac{1}{5}$ ← while in the Alps, 1 out of 5 spend a day walking → **WALKING**

$\frac{4}{5}$ ← while in the Alps, 4 out of 5 spend a day cycling → **CYCLING**

PARIS

THE ALPS

STAGE 2: WHERE TO STAY?

STAGE 3: DAY TRIPS

$\frac{1}{3}$ ← while in Rome, 1 out of 3 visit the Colosseum → **COLOSSEUM**

$\frac{2}{3}$ ← while in Rome, 2 out of 3 visit the Vatican → **VATICAN**

$\frac{3}{4}$ ← while in Naples, 3 out of 4 visit Pompeii → **POMPEII**

$\frac{1}{4}$ ← while in Naples, 1 out of 4 visit Vesuvius → **VESUVIUS**

ROME

NAPLES

△ **Stage 2 of multiple events**
Stage 2 shows the chances of a randomly selected person staying in a specific location.

△ **Stage 3 of multiple events**
Stage 3 shows the chances of a randomly selected person making a specific day trip.

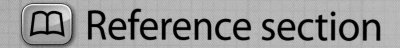

Mathematical signs and symbols

This table shows a selection of signs and symbols commonly used in mathematics. Using signs and symbols, mathematicians can express complex equations and formulas in a standardized way that is universally understood.

Symbol	Definition	Symbol	Definition	Symbol	Definition
$+$	plus; positive	$:$	ratio of (6:4)	∞	infinity
$-$	minus; negative	$::$	proportionately equal (1:2∷2:4)	n^2	squared number
$\pm$	plus or minus; positive or negative; degree of accuracy	$\approx, \doteq, \triangleq$	approximately equal to; equivalent to; similar to	n^3	cubed number
				n^4, n^5, etc	power, index
$\mp$	minus or plus; negative or positive	$\cong$	congruent to; identical with	$\sqrt[2]{}$	square root
$\times$	multiplied by (6 × 4)	$>$	greater than	$\sqrt[3]{}, \sqrt[4]{}$	cube root, fourth root, etc.
$\cdot$	multiplied by (6·4); scalar product of two vectors (A·B)	$\gg$	much greater than	$\%$	per cent
		$\ngtr$	not greater than	$°$	degrees (°C); degree of arc, for example 90°
$\div$	divided by (6 ÷ 4)	$<$	less than	$\angle, \angle^s$	angle(s)
$/$	divided by; ratio of ($^6/_4$)	$\ll$	much less than	$\underline{\vee}$	equiangular
$-$	divided by; ratio of ($\frac{6}{4}$)	$\nless$	not less than	π	(pi) the ratio of the circumference to the diameter of a circle = 3.14
$\bigcirc$	circle	$\geqslant, \geqq, \gtreqqless$	equal to or greater than		
$\blacktriangle$	triangle	$\leqslant, \leqq, \lesseqqgtr$	equal to or less than		
$\square$	square	$\propto$	directly proportional to	α	alpha (unknown angle)
$\square$	rectangle	$(\,)$	parentheses, can mean multiply	θ	theta (unknown angle)
$\square$	parallelogram			$\perp$	perpendicular
$=$	equals	$-$	vinculum: division (a-b); chord of circle or length of line (AB);	$\llcorner$	right angle
$\neq$	not equal to			$\parallel, \Rightarrow$	parallel
$\equiv$	identical with; congruent to	$\overrightarrow{AB}$	vector	$\therefore$	therefore
$\not\equiv$	not identical with	$\overline{AB}$	line segment	$\because$	because
$\triangleq$	corresponds to	$\overleftrightarrow{AB}$	line	$\underline{m}$	measured by

Prime numbers

A prime number is any number that can only be exactly divided by 1 and itself without leaving a remainder. By definition, 1 is not a prime. There is no one formula for yielding every prime. Shown here are the first 250 prime numbers.

2	3	5	7	11	13	17	19	23	29
31	37	41	43	47	53	59	61	67	71
73	79	83	89	97	101	103	107	109	113
127	131	137	139	149	151	157	163	167	173
179	181	191	193	197	199	211	223	227	229
233	239	241	251	257	263	269	271	277	281
283	293	307	311	313	317	331	337	347	349
353	359	367	373	379	383	389	397	401	409
419	421	431	433	439	443	449	457	461	463
467	479	487	491	499	503	509	521	523	541
547	557	563	569	571	577	587	593	599	601
607	613	617	619	631	641	643	647	653	659
661	673	677	683	691	701	709	719	727	733
739	743	751	757	761	769	773	787	797	809
811	821	823	827	829	839	853	857	859	863
877	881	883	887	907	911	919	929	937	941
947	953	967	971	977	983	991	997	1009	1013
1019	1021	1031	1033	1039	1049	1051	1061	1063	1069
1087	1091	1093	1097	1103	1109	1117	1123	1129	1151
1153	1163	1171	1181	1187	1193	1201	1213	1217	1223
1229	1231	1237	1249	1259	1277	1279	1283	1289	1291
1297	1301	1303	1307	1319	1321	1327	1361	1367	1373
1381	1399	1409	1423	1427	1429	1433	1439	1447	1451
1453	1459	1471	1481	1483	1487	1489	1493	1499	1511
1523	1531	1543	1549	1553	1559	1567	1571	1579	1583

Squares, cubes, and roots

The table below shows the square, cube, square root, and cube root of whole numbers, to 3 decimal places.

No.	Square	Cube	Square root	Cube root
1	1	1	1.000	1.000
2	4	8	1.414	1.260
3	9	27	1.732	1.442
4	16	64	2.000	1.587
5	25	125	2.236	1.710
6	36	216	2.449	1.817
7	49	343	2.646	1.913
8	64	512	2.828	2.000
9	81	729	3.000	2.080
10	100	1,000	3.162	2.154
11	121	1,331	3.317	2.224
12	144	1,728	3.464	2.289
13	169	2,197	3.606	2.351
14	196	2,744	3.742	2.410
15	225	3,375	3.873	2.466
16	256	4,096	4.000	2.520
17	289	4,913	4.123	2.571
18	324	5,832	4.243	2.621
19	361	6,859	4.359	2.668
20	400	8,000	4.472	2.714
25	625	15,625	5.000	2.924
30	900	27,000	5.477	3.107
50	2,500	125,000	7.071	3.684

Multiplication table

This multiplication table shows the products of each whole number from 1 to 12, multiplied by each whole number from 1 to 12.

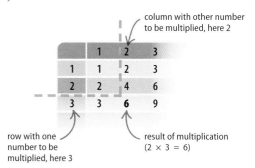

column with other number to be multiplied, here 2

row with one number to be multiplied, here 3

result of multiplication (2 × 3 = 6)

	1	2	3	4	5	6	7	8	9	10	11	12
1	1	2	3	4	5	6	7	8	9	10	11	12
2	2	4	6	8	10	12	14	16	18	20	22	24
3	3	6	9	12	15	18	21	24	27	30	33	36
4	4	8	12	16	20	24	28	32	36	40	44	48
5	5	10	15	20	25	30	35	40	45	50	55	60
6	6	12	18	24	30	36	42	48	54	60	66	72
7	7	14	21	28	35	42	49	56	63	70	77	84
8	8	16	24	32	40	48	56	64	72	80	88	96
9	9	18	27	36	45	54	63	72	81	90	99	108
10	10	20	30	40	50	60	70	80	90	100	110	120
11	11	22	33	44	55	66	77	88	99	110	121	132
12	12	24	36	48	60	72	84	96	108	120	132	144

Units of measurement

A unit of measurement is a quantity used as a standard, allowing values of things to be compared. These include seconds (time), metres (length), and kilograms (mass). Two widely used systems of measurement are the metric system and the imperial system.

AREA		
metric		
100 square millimetres (mm^2)	=	1 square centimetre (cm^2)
10,000 square centimetres (cm^2)	=	1 square metre (m^2)
10,000 square metres (m^2)	=	1 hectare (ha)
100 hectares (ha)	=	1 square kilometre (km^2)
1 square kilometre (km^2)	=	1,000,000 square metres (m^2)
imperial		
144 square inches (sq in)	=	1 square foot (sq ft)
9 square feet (sq ft)	=	1 square yard (sq yd)
1,296 square inches (sq in)	=	1 square yard (sq yd)
43,560 square feet (sq ft)	=	1 acre
640 acres	=	1 square mile (sq mile)

LIQUID VOLUME		
metric		
1,000 millilitres (ml)	=	1 litre (l)
100 litres (l)	=	1 hectolitre (hl)
10 hectolitres (hl)	=	1 kilolitre (kl)
1,000 litres (l)	=	1 kilolitre (kl)
imperial		
8 fluid ounces (fl oz)	=	1 cup
20 fluid ounces (fl oz)	=	1 pint (pt)
4 gills (gi)	=	1 pint (pt)
2 pints (pt)	=	1 quart (qt)
4 quarts (qt)	=	1 gallon (gal)
8 pints (pt)	=	1 gallon (gal)

MASS		
metric		
1,000 milligrams (mg)	=	1 gram (g)
1,000 grams (g)	=	1 kilogram (kg)
1,000 kilograms (kg)	=	1 tonne (t)
imperial		
16 ounces (oz)	=	1 pound (lb)
14 pounds (lb)	=	1 stone
112 pounds (lb)	=	1 hundredweight
20 hundredweight	=	1 ton

LENGTH		
metric		
10 millimetres (mm)	=	1 centimetre (cm)
100 centimetres (cm)	=	1 metre (m)
1,000 millimetres (mm)	=	1 metre (m)
1,000 metres (m)	=	1 kilometre (km)
imperial		
12 inches (in)	=	1 foot (ft)
3 feet (ft)	=	1 yard (yd)
1,760 yards (yd)	=	1 mile
5,280 feet (ft)	=	1 mile
8 furlongs	=	1 mile

TIME		
metric and imperial		
60 seconds	=	1 minute
60 minutes	=	1 hour
24 hours	=	1 day
7 days	=	1 week
52 weeks	=	1 year
1 year	=	12 months

TEMPERATURE		Fahrenheit	Celsius	Kelvin
Boiling point of water	=	212°	100°	373°
Freezing point of water	=	32°	0°	273°
Absolute zero	=	−459°	−273°	0°

Conversion tables

The tables below show metric and imperial equivalents for common measurements for length, area, mass, and volume. Conversions between Celcius, Fahrenheit, and Kelvin temperature require formulas, which are also given below.

LENGTH		
metric		**imperial**
1 millimetre (mm)	=	0.03937 inch (in)
1 centimetre (cm)	=	0.3937 inch (in)
1 metre (m)	=	1.0936 yards (yd)
1 kilometre (km)	=	0.6214 mile
imperial		**metric**
1 inch (in)	=	2.54 centimetres (cm)
1 foot (ft)	=	0.3048 metre (m)
1 yard (yd)	=	0.9144 metre (m)
1 mile	=	1.6093 kilometres (km)
1 nautical mile	=	1.853 kilometres (km)

AREA		
metric		**imperial**
1 square centimetre (cm^2)	=	0.155 square inch (sq in)
1 square metre (m^2)	=	1.196 square yard (sq yd)
1 hectare (ha)	=	2.4711 acres
1 square kilometre (km^2)	=	0.3861 square miles
imperial		**metric**
1 square inch (sq in)	=	6.4516 square centimetres (cm^2)
1 square foot (sq ft)	=	0.0929 square metre (m^2)
1 square yard (sq yd)	=	0.8361 square metre (m^2)
1 acre	=	0.4047 hectare (ha)
1 square mile	=	2.59 square kilometres (km^2)

MASS		
metric		**imperial**
1 milligram (mg)	=	0.0154 grain
1 gram (g)	=	0.0353 ounce (oz)
1 kilogram (kg)	=	2.2046 pounds (lb)
1 tonne/metric ton (t)	=	0.9842 imperial ton
imperial		**metric**
1 ounce (oz)	=	28.35 grams (g)
1 pound (lb)	=	0.4536 kilogram (kg)
1 stone	=	6.3503 kilogram (kg)
1 hundredweight (cwt)	=	50.802 kilogram (kg)
1 imperial ton	=	1.016 tonnes/metric tons

VOLUME		
metric		**imperial**
1 cubic centimetre (cm^3)	=	0.061 cubic inch (in^3)
1 cubic decimetre (dm^3)	=	0.0353 cubic foot (ft^3)
1 cubic metre (m^3)	=	1.308 cubic yard (yd^3)
1 litre (l)/1 dm^3	=	1.76 pints (pt)
1 hectolitre (hl)/100 l	=	21.997 gallons (gal)
imperial		**metric**
1 cubic inch (in^3)	=	16.387 cubic centimetres (cm^3)
1 cubic foot (ft^3)	=	0.0283 cubic metres (m^3)
1 fluid ounce (fl oz)	=	28.413 millilitres (ml)
1 pint (pt)/20 fl oz	=	0.5683 litre (l)
1 gallon/8 pt	=	4.5461 litres (l)

TEMPERATURE		
To convert from Fahrenheit (°F) to Celsius (°C)	=	C = (F − 32) × 5 ÷ 9
To convert from Celsius (°C) Fahrenheit (°F)	=	F = (C × 9 ÷ 5) + 32
To convert from Celsius (°C) to Kelvin (K)	=	K = C + 273
To convert from Kelvin (K) to Celsius (°C)	=	C = K − 273

Fahrenheit °F	−4	14	32	50	68	86	104	122	140	158	176	194	212
Celsius °C	−20	−10	0	10	20	30	40	50	60	70	80	90	100
Kelvin	253	263	273	283	293	303	313	325	333	343	353	363	373

How to convert

The table below shows how to convert between metric and British imperial units of measurement. The left table shows how to convert from one unit to its metric or imperial equivalent. The right table shows how to do the reverse conversion.

HOW TO CONVERT METRIC and IMPERIAL MEASURES		
to change	to	multiply by
acres	hectares	0.4047
centimetres	feet	0.03281
centimetres	inches	0.3937
cubic centimetres	cubic inches	0.061
cubic feet	cubic metres	0.0283
cubic inches	cubic centimetres	16.3871
cubic metres	cubic feet	35.315
feet	centimetres	30.48
feet	metres	0.3048
gallons	litres	4.546
grams	ounces	0.0353
hectares	acres	2.471
inches	centimetres	2.54
kilograms	pounds	2.2046
kilometres	miles	0.6214
kilometres per hour	miles per hour	0.6214
litres	gallons	0.2199
litres	pints	1.7598
metres	feet	3.2808
metres	yards	1.0936
metres per minute	centimetres per second	1.6667
metres per minute	feet per second	0.0547
miles	kilometres	1.6093
miles per hour	kilometres per hour	1.6093
miles per hour	metres per second	0.447
millimetres	inches	0.0394
ounces	grams	28.3495
pints	litres	0.5682
pounds	kilograms	0.4536
square centimetres	square inches	0.155
square inches	square centimetres	6.4516
square feet	square metres	0.0929
square kilometres	square miles	0.386
square metres	square feet	10.764
square metres	square yards	1.196
square miles	square kilometres	2.5899
square yards	square metres	0.8361
tonnes (metric)	tons (imperial)	0.9842
tons (imperial)	tonnes (metric)	1.0216
yards	metres	0.9144

HOW TO CONVERT METRIC and IMPERIAL MEASURES		
to change	to	divide by
hectares	acres	0.4047
feet	centimetres	0.03281
inches	centimetres	0.3937
cubic inches	cubic centimetres	0.061
cubic metres	cubic feet	0.0283
cubic centimetres	cubic inches	16.3871
cubic feet	cubic metres	35.315
centimetres	feet	30.48
metres	feet	0.3048
litres	gallons	4.546
ounces	grams	0.0353
acres	hectares	2.471
centimetres	inches	2.54
pounds	kilograms	2.2046
miles	kilometres	0.6214
miles per hour	kilometres per hour	0.6214
gallons	litres	0.2199
pints	litres	1.7598
feet	metres	3.2808
yards	metres	1.0936
centimetres per second	metres per minute	1.6667
feet per second	metres per minute	0.0547
kilometres	miles	1.6093
kilometres per hour	miles per hour	1.6093
metres per second	miles per hour	0.447
inches	millimetres	0.0394
grams	ounces	28.3495
litres	pints	0.5682
kilograms	pounds	0.4536
square inches	square centimetres	0.155
square centimetres	square inches	6.4516
square metres	square feet	0.0929
square miles	square kilometres	0.386
square feet	square metres	10.764
square yards	square metres	1.196
square kilometres	square miles	2.5899
square metres	square yards	0.8361
tons (imperial)	tonnes (metric)	0.9842
tonnes (metric)	tons (imperial)	1.0216
metres	yards	0.9144

Numerical equivalents

Percentages, decimals, and fractions are different ways of presenting a numerical value as a proportion of a given amount. For example, 10% (10 per cent) has the equivalent value of the decimal 0.1 and the fraction 1/10.

%	Decimal	Fraction	%	Decimal	Fraction	%	Decimal	Fraction	%	Decimal	Fraction	%	Decimal	Fraction
1	0.01	1/100	12.5	0.125	1/8	24	0.24	6/25	36	0.36	9/25	49	0.49	49/100
2	0.02	1/50	13	0.13	13/100	25	0.25	1/4	37	0.37	37/100	50	0.5	1/2
3	0.03	3/100	14	0.14	7/50	26	0.26	13/50	38	0.38	19/50	55	0.55	11/20
4	0.04	1/25	15	0.15	3/20	27	0.27	27/100	39	0.39	39/100	60	0.6	3/5
5	0.05	1/20	16	0.16	4/25	28	0.28	7/25	40	0.4	2/5	65	0.65	13/20
6	0.06	3/50	16.66	0.166	1/6	29	0.29	29/100	41	0.41	41/100	66.66	0.666	2/3
7	0.07	7/100	17	0.17	17/100	30	0.3	3/10	42	0.42	21/50	70	0.7	7/10
8	0.08	2/25	18	0.18	9/50	31	0.31	31/100	43	0.43	43/100	75	0.75	3/4
8.33	0.083	1/12	19	0.19	19/100	32	0.32	8/25	44	0.44	11/25	80	0.8	4/5
9	0.09	9/100	20	0.2	1/5	33	0.33	33/100	45	0.45	9/20	85	0.85	17/20
10	0.1	1/10	21	0.21	21/100	33.33	0.333	1/3	46	0.46	23/50	90	0.9	9/10
11	0.11	11/100	22	0.22	11/50	34	0.34	17/50	47	0.47	47/100	95	0.95	19/20
12	0.12	3/25	23	0.23	23/100	35	0.35	7/20	48	0.48	12/25	100	1.00	1

Angles

An angle shows the amount that a line "turns" as it extends in a direction away from a fixed point.

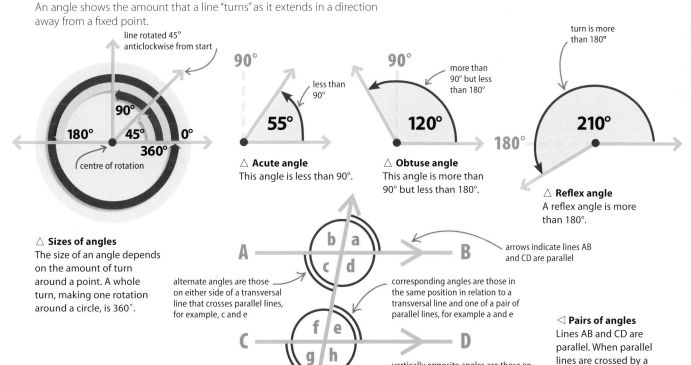

line rotated 45°
anticlockwise from start

90°

45°

180°

0°

360°

centre of rotation

90°

less than 90°

55°

△ **Acute angle**
This angle is less than 90°.

90°

more than 90° but less than 180°

120°

△ **Obtuse angle**
This angle is more than 90° but less than 180°.

turn is more than 180°

210°

180°

△ **Reflex angle**
A reflex angle is more than 180°.

△ **Sizes of angles**
The size of an angle depends on the amount of turn around a point. A whole turn, making one rotation around a circle, is 360°.

A
b a
c d
B

alternate angles are those on either side of a transversal line that crosses parallel lines, for example, c and e

corresponding angles are those in the same position in relation to a transversal line and one of a pair of parallel lines, for example a and e

arrows indicate lines AB and CD are parallel

C
f e
g h
D

transversal line crosses parallel lines

vertically opposite angles are those on opposite sides of a point where two lines cross, for example, f and h

◁ **Pairs of angles**
Lines AB and CD are parallel. When parallel lines are crossed by a transversal, pairs of equal angles are created.

Shapes

Two-dimensional shapes with straight lines are called polygons. They are named according to the number of sides they have. The number of sides is also equal to the number of interior angles. A circle has no straight lines, so it is not a polygon, although it is a two-dimensional shape.

△ **Circle**
A shape formed by a curved line that is always the same distance from a central point.

△ **Triangle**
A polygon with three sides and three interior angles.

△ **Quadrilateral**
A polygon with four sides and four interior angles.

△ **Square**
A quadrilateral with four equal sides and four equal interior angles of 90˚ (right angles).

△ **Rectangle**
A quadrilateral with four equal interior angles and opposite sides of equal length.

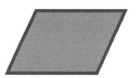

△ **Parallelogram**
A quadrilateral with two pairs of parallel sides and opposite sides of equal length.

△ **Pentagon**
A polygon with five sides and five interior angles.

△ **Hexagon**
A polygon with six sides and six interior angles.

△ **Heptagon**
A polygon with seven sides and seven interior angles.

△ **Nonagon**
A polygon with nine sides and nine interior angles.

△ **Decagon**
A polygon with ten sides and ten interior angles.

△ **Hendecagon**
A polygon with eleven sides and eleven interior angles.

Sequences

A sequence is a series of numbers written as an ordered list where there is a particular pattern or "rule" that relates each number in the list to the numbers before and after it. Examples of important mathematical sequences are shown below.

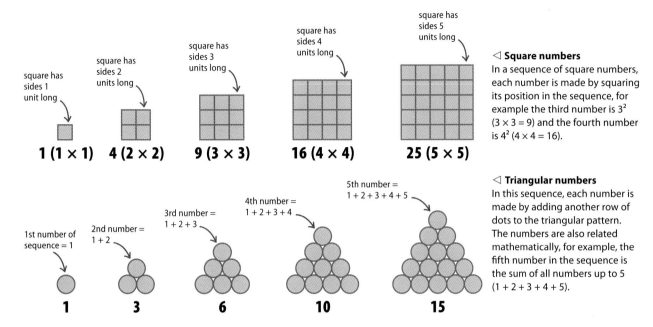

◁ **Square numbers**
In a sequence of square numbers, each number is made by squaring its position in the sequence, for example the third number is 3^2 ($3 \times 3 = 9$) and the fourth number is 4^2 ($4 \times 4 = 16$).

◁ **Triangular numbers**
In this sequence, each number is made by adding another row of dots to the triangular pattern. The numbers are also related mathematically, for example, the fifth number in the sequence is the sum of all numbers up to 5 ($1 + 2 + 3 + 4 + 5$).

Fibonacci sequence

Named after the Italian mathematician Leonardo Fibonacci (c.1175–c.1250), the Fibonacci sequence starts with 1. The second number is also 1. After that, each number in the sequence is the sum of the two numbers before it, for example, the sixth number, 8, is the sum of the fourth and fifth numbers, 3 and 5 ($3 + 5 = 8$).

Pascal's Triangle

Pascal's triangle is a triangular arrangement of numbers. The number at the top of the triangle is 1, and every number down each side is also 1. Each of the other numbers is the sum of the two numbers diagonally above it; for example, in the third row, the 2 is made by adding the two 1s in the row above.

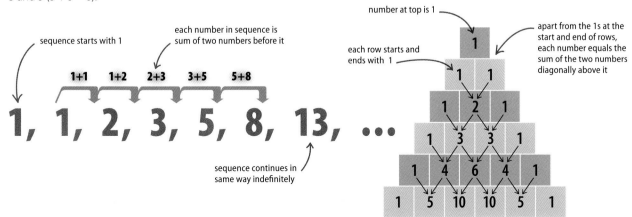

FORMULAS

Formulas are mathematical "recipes" that relate various quantities or terms, so that if the value of one is unknown, it can be worked out if the values of the other terms in the formula are known.

Interest

There are two types of interest – simple and compound. In simple interest, the interest is paid only on the capital. In compound interest, the interest itself earns interest.

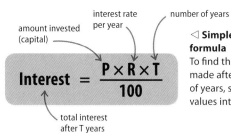

amount invested (capital)

interest rate per year

number of years

$$\text{Interest} = \frac{P \times R \times T}{100}$$

total interest after T years

◁ **Simple interest formula**
To find the simple interest made after a given number of years, substitute real values into this formula.

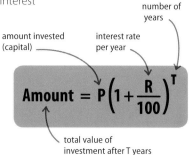

amount invested (capital)

interest rate per year

number of years

$$\text{Amount} = P\left(1 + \frac{R}{100}\right)^T$$

total value of investment after T years

◁ **Compound interest formula**
To find the total value of an investment (capital + interest) after a given number of years, substitute real values into this formula.

Formulas in algebra

Algebra is the branch of mathematics that uses symbols to represent numbers and the relationship between them. Useful formulas are the standard formula of a quadratic equation and the formula for solving it.

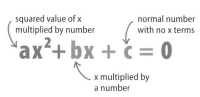

squared value of x multiplied by number

normal number with no x terms

$$ax^2 + bx + c = 0$$

x multiplied by a number

△ **Quadratic equation**
Quadratic equations take the form shown above. They can be solved by the quadratic formula.

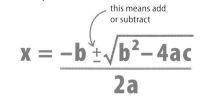

this means add or subtract

$$x = \frac{-b \pm \sqrt{b^2 - 4ac}}{2a}$$

△ **The quadratic formula**
This formula can be used to solve any quadratic equation. There are always two solutions.

symbol for pi

value to 2 decimal places

$$\pi = 3.14$$

value to 20 decimal places

$$3.14159265358979323846$$

◁ **The value of pi**
Pi occurs in many formulas, such as the formula used for working out the area of a circle. The numbers after the decimal point in pi go on for ever and do not follow any pattern.

Formulas in trigonometry

Three of the most useful formulas in trigonometry are those used to find the unknown angles of a right-angled triangle when two of its sides are known.

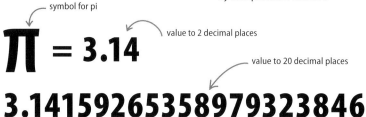

$$\sin A = \frac{\text{opposite}}{\text{hypotenuse}}$$

$$\cos A = \frac{\text{adjacent}}{\text{hypotenuse}}$$

$$\tan A = \frac{\text{opposite}}{\text{adjacent}}$$

△ **The sine formula**
This formula is used to find the size of an angle when the side opposite the angle and the hypotenuse are known.

△ **The cosine formula**
This formula is used to find the size of an angle when the side adjacent to the angle and the hypotenuse are known.

△ **The tangent formula**
This formula is used to find the size of an angle when the sides opposite and adjacent to the angle are known.

Area

The area of a shape is the amount of space inside it. Formulas for working out the areas of common shapes are given below.

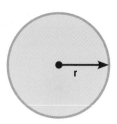

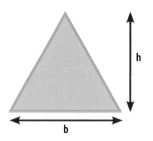

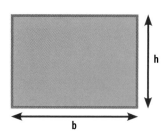

$$\text{area} = \pi r^2$$

$$\text{area} = \frac{1}{2}\text{bh}$$

$$\text{area} = \text{bh}$$

△ **Circle**
The area of a circle equals pi ($\pi = 3.14$) multiplied by the square of its radius.

△ **Triangle**
The area of a triangle equals half multiplied by its base multiplied by its vertical height.

△ **Rectangle**
The area of a rectangle equals its base multiplied by its height.

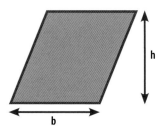

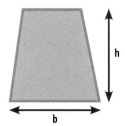

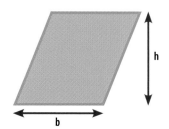

$$\text{area} = \text{bh}$$

$$\text{area} = \frac{1}{2}\text{h}(\text{b}_1+\text{b}_2)$$

$$\text{area} = \text{bh}$$

△ **Parallelogram**
The area of a parallelogram equals its base multiplied by its vertical height.

△ **Trapezium**
The area of a trapezium equals the sum of the two parallel sides, multiplied by the vertical height, then divided by 2.

△ **Rhombus**
The area of a rhombus equals its base multiplied by its vertical height.

Pythagoras' theorem

This theorem relates the lengths of all the sides of a right-angled triangle, so that if any two sides are known, the length of the third side can be worked out.

side a   side c (hypotenuse)

$$a^2 + b^2 = c^2$$

side b

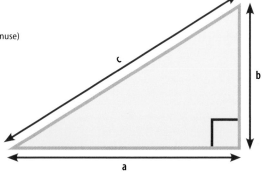

◁ **The theorem**
In a right-angled triangle the square of the hypotenuse (the largest side, c) is the sum of the squares of the other two sides (a and b).

Surface and volume area

The illustrations below show three-dimensional shapes and the formulas for calculating their surface areas and their volumes. In the formulas, two letters together means that they are multiplied together, for example "2r" means "2" multiplied by "r". Pi (Π) is 3.14, (to 2 decimal places).

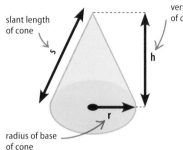

slant length of cone

s

vertical height of cone

h

r

radius of base of cone

◁ **Cone**
The surface area of a cone can be found from the radius of its base, its vertical height, and its slant length. The volume can be found from the height and radius.

$$\text{surface area} = \pi rs + \pi r^2$$
$$\text{volume} = \frac{1}{3}\pi r^2 h$$

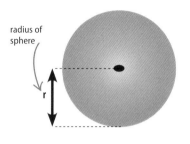

radius of sphere

r

◁ **Sphere**
The surface area and volume of a sphere can be found when only its radius is known, as pi is a constant number (equal to 3.14, to 2 decimal places).

$$\text{surface area} = 4\pi r^2$$
$$\text{volume} = \frac{4}{3}\pi r^3$$

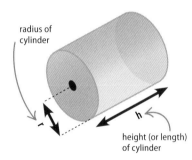

radius of cylinder

r

h

height (or length) of cylinder

◁ **Cylinder**
The surface area and volume of a cylinder can be found from its radius and height (or length).

$$\text{surface area} = 2\pi r\,(h+r)$$
$$\text{volume} = \pi r^2 h$$

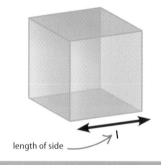

l

length of side

◁ **Cube**
The surface area and volume of a cube can be found when only the length of its sides is known.

$$\text{surface area} = 6l^2$$
$$\text{volume} = l^3$$

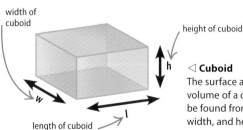

width of cuboid

w

height of cuboid

h

l

length of cuboid

◁ **Cuboid**
The surface area and volume of a cuboid can be found from its length, width, and height.

$$\text{surface area} = 2(lh+lw+hw)$$
$$\text{volume} = lwh$$

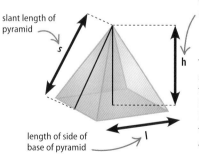

slant length of pyramid

s

vertical height of pyramid

h

l

length of side of base of pyramid

◁ **Square pyramid**
The surface area of a square pyramid can be found from the slant length and the side of its base. Its volume can be found from its height and the side of its base.

$$\text{surface area} = 2ls+l^2$$
$$\text{volume} = \frac{1}{3}l^2 h$$

Parts of a circle

Various properties of a circle can be measured using certain characteristics, such as the radius, circumference, or length of an arc, with the formulas given below. Pi (∏) is the ratio of the circumference to the diameter of a circle; pi is equal to 3.14 (to 2 decimal places).

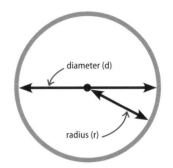

◁ **Diameter and radius**
The diameter of a circle is a straight line running right across the circle and through its centre. It is twice the length of the radius (the line from the centre to the circumference).

$$\text{diameter} = 2r$$

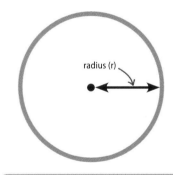

◁ **Diameter and circumference**
The diameter of a circle can be found when only its circumference (the distance around the edge) is known.

$$\text{diameter} = \frac{c}{\pi}$$

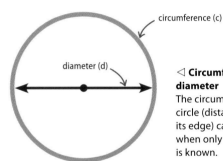

◁ **Circumference and diameter**
The circumference of a circle (distance around its edge) can be found when only its diameter is known.

$$\text{circumference} = \pi d$$

◁ **Circumference and radius**
The circumference of a circle (distance around its edge) can be found when only its radius is known.

$$\text{circumference} = 2\pi r$$

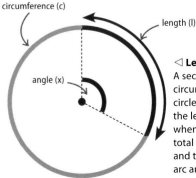

◁ **Length of an arc**
A section of the circumference of a circle is known as an arc, the length can be found when the circle's total circumference and the angle of the arc are known.

$$\text{length of an arc} = \frac{x}{360} \times c$$

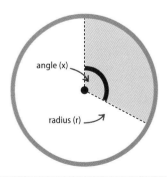

◁ **Area of a sector**
The area of a sector (or "slice") of a circle can be found when the circle's area and the angle of the sector are known.

$$\text{area of a sector} = \frac{x}{360} \times \pi r^2$$

 Glossary

Acute
An acute angle is an angle that is smaller than 90°.

Addition
Working out the sum of a group of numbers. Addition is represented by the + symbol, e.g. 2 + 3 = 5. The order the numbers are added in does not affect the answer: 2 + 3 = 3 + 2.

Adjacent
A term meaning "next to". In two-dimensional shapes two sides are adjacent if they are next to each other and meet at the same point (vertex). Two angles are adjacent if they share a vertex and a side.

Algebra
The use of letters or symbols in place of unknown numbers to generalize the relationship between them.

Alternate angle
Alternate angles are formed when two parallel lines are crossed by another straight line. They are the angles on the opposite sides of each of the lines. Alternate angles are equal.

Angle
The amount of turn between two lines that meet at a point. Angles are measured in degrees, for example, 45°.

Anticlockwise
Movement in the opposite direction to that of a clock's hand.

Apex
The tip of something e.g. the vertex of a cone.

Arc
A curve that is part of the circumference of a circle.

Area
The amount of space within a two-dimensional outline. Area is measured in units squared, e.g. cm^2.

Arithmetic
Calculations involving addition, subtraction, multiplication, division, or combinations of these.

Average
The typical value of a group of numbers. There are three types of average: median, mode, and mean.

Axis (plural: axes)
Reference lines used in graphs to define coordinates and measure distances. The horizontal axis is the x-axis, the vertical axis is the y-axis.

Balance
Equality on every side, so that there is no unequal weighting, e.g. in an equation, the left-hand side of the equals sign must balance with the right-hand side.

Bar chart
A graph where quantities are represented by rectangles (bars), which are the same width but varying heights. A greater height means a greater amount.

Base
The base of a shape is its bottom edge. The base of a three-dimensional object is its bottom face.

Bearing
A compass reading. The angle measured clockwise from the North direction to the target direction, and given as 3 figures.

Bisect
To divide into two equal halves, e.g. to bisect an angle or a line.

Box-and-whisker diagram
A way to represent statistical data. The box is constructed from lines indicating where the lower quartile, median, and upper quartile measurements fall on a graph, and the whiskers mark the upper and lower limits of the range.

Brackets
1. Brackets indicate the order in which calculations must be done – calculations in brackets must be done first e.g. 2 x (4 +1) = 10.
2. Brackets mark a pair of numbers that are coordinates, e.g. (1, 1).
3. When a number appears before a bracketed calculation it means that the result of that calculation must be multiplied by that number.

Break even
In order to break even a business must earn as much money as it spends. At this point revenue and costs are equal.

Calculator
An electronic tool used to solve arithmetic.

Chart
An easy-to-read visual representation of data, such as a graph, table, or map.

Chord
A line that connects two different points on a curve, often on the circumference of a circle.

Circle
A round shape with only one edge, which is a constant distance from the centre point.

Circumference
The edge of a circle.

Clockwise
A direction the same as that of a clock's hand.

Coefficient
The number in front of a letter in algebra. In the equation $x^2 + 5x + 6 = 0$ the coefficient of 5x is 5.

Common factor
A common factor of two or more numbers divides exactly into each of those numbers, e.g. 3 is a common factor of 6 and 18.

Compass
1. A magnetic instrument that shows the position of North and allows bearings to be found.
2. A tool that holds a pencil in a fixed position, allowing circles and arcs to be drawn.

Composite number
A number with more than two factors. A number is composite if it is not a prime number e.g. 4 is a composite factor as it has 1, 2, and 4 as factors.

Concave
Something curving inwards. A polygon is concave if one of its interior angles is greater than 180°.

Cone
A three-dimensional object with a circular base and a single point at its top.

Congruent/congruence
Two shapes are congruent if they are both the same shape and size.

Constant
A quantity that does not change and so has a fixed value, e.g. in the equation y = x + 2, the number 2 is a constant.

Construction
The drawing of shapes in geometry accurately, often with the aid of a compass and ruler.

Conversion
The change from one set of units to another e.g. the conversion from miles into kilometres.

Convex
Something curving ourwards. A polygon is convex if all its interior angles are less than 180°.

Coordinate
Coordinates show the position of points on a graph or map, and are written in the form (x,y), where x is the horizontal position and y is the vertical position.

Correlate/correlation
There is a correlation between two things if a change in one causes a change in the other.

Corresponding angles
Corresponding angles are formed when two parallel lines are crossed by another straight side. They are the angles in the same position i.e. on the same side of each of the lines. Corresponding angles are equal.

Cosine
In trigonometry, cosine is the ratio of the side adjacent to a given angle with the hypotenuse of a right-angled triangle.

Cross section
A two-dimensional slice of a three-dimensional object.

Cube
A three-dimensional object made up of 6 identical square faces, 8 vertices, and 12 edges.

Cube root
A number's cube root is the number which, multiplied by itself twice, equals the given number. A cube root is indicated by this sign $\sqrt[3]{}$

Cubed number
Cubing a number means multiplying it by itself twice e.g. 8 is a cubed number because $2 \times 2 \times 2 = 8$, or 2^3.

Cuboids
A three-dimensional object made of 6 faces (2 squares at opposite ends with 4 rectangles between), 8 vertices, and 12 edges.

Currency
A system of money within a country e.g. the currency in the US is $.

Curve
A line that bends smoothly. A quadratic equation represented on a graph is also a curve.

Cyclic quadrilateral
A shape with 4 vertices and 4 edges, and where every vertex is on the circumference of a circle.

Cylinder
A three-dimensional object with two parallel, congruent circles at opposite ends.

Data
A set of information, e.g. a collection of numbers or measurements.

Debit
An amount of money spent and removed from an account.

Debt
An amount of money that has been borrowed, and is therefore owed.

Decimal
1. A number system based on 10 (using the digits 0, 1, 2, 3, 4, 5, 6, 7, 8, and 9).
2. A number containing a decimal place.

Decimal point
The dot between the whole part of a number and the fractional part e.g. 2.5.

Decimal place
The position of the digit after the decimal point.

Degrees
The unit of measurement of an angle, represented by the symbol °.

Denominator
The number on the bottom of a fraction e.g. 3 is the denominator of $^2/_3$.

Density
The amount of mass per unit of volume, i.e. density = mass ÷ volume.

Diagonal
A line that joins two vertices of a shape or object that are not adjacent to each other.

Diameter
A straight line touching two points on the edge of a circle and passing through the centre.

Difference
The amount by which one quantity is bigger or smaller than another quantity.

Digit
A single number, e.g. 34 is made up of the digits 3 and 4.

Dimension
The directions in which measurements can be made e.g. a solid object has three dimensions: its length, height, and width.

Direct proportion
Two numbers are in direct proportion if they increase or decrease proportionately, e.g. doubling one of them means the other also doubles.

Distribution
In probability and statistics, the distribution gives the range of values unidentified random variables can take and their probabilities.

Division/divide
The splitting of a number into equal parts. Division is shown by the symbol ÷ e.g. 12 ÷ 3 = 4 or by / as used in fractions, e.g. $^2/_3$.

Double negative
Two negative signs together create a double negative, which then becomes equal to a positive e.g. 5 − (−2) = 5 + 2.

Enlargement
The process of making something bigger, such as a transformation, where everything is multiplied by the same amount.

Equal
Things of the same value are equal, shown by the equals sign, =.

Equation
A mathematical statement that things are equal.

Equiangular
A shape is equiangular if all its angles are equal.

Equidistant
A point is equidistant to two or more points if it is the same distance from them.

Equilateral triangle
A triangle that has three 60° angles and sides of equal length.

Equiprobable events
Two events are equiprobable if they are equally likely to happen.

Equivalent fractions
Fractions that are equal but have different numerators and denominators e.g. $^1/_2$, $^2/_4$, and $^5/_{10}$ are equivalent fractions.

Estimation
An approximated amount or an approximation the answer to a calculation, often made by rounding up or down.

Even number
A number that is divisible by 2 e.g. -18, -6, 0, 2.

Exchange rate
The exchange rate describes what an amount of one currency is valued at in another currency.

Exponent
See power

Expression
A combination of numbers, symbols, and unknown variables that does not contain an equal sign.

Exterior angle
1. An angle formed on the outside of a polygon, when one side is extended outwards.
2. The angles formed in the region outside two lines intersected by another line.

Faces
The flat surfaces of a three-dimensional object, bordered by edges.

Factor
A number that divides exactly into another, larger number, e.g. 2 and 5 are both factors of 10.

Factorization/factorize
1. Rewriting a number as the multiplication of its factors, e.g. $12 = 2 \times 2 \times 3$.
2. Rewriting an expression as the multiplication of smaller expressions e.g. $x^2 + 5x + 6 = (x + 2)(x + 3)$.

Fibonacci sequence
A sequence formed by adding the previous two numbers in the sequence together, which begins with 1, 1. The first ten numbers in the sequence are 1, 1, 2, 3, 5, 8, 13, 21, 34, and 55.

Formula
A rule that describes the relationship between variables, and is usually written as symbols, e.g. the formula for calculating the area of a circle is $A = 2\pi r$, in which A represents the area and r is the radius.

Fraction
A part of an amount, represented by one number (the numerator) on top of another number (the denominator) e.g. $^2/_3$.

Frequency
1. The number of times something occurs during a fixed period of time.
2. In statistics, the number of individuals in a class.

Geometry
The mathematics of shapes. Looks at the relationships between points, lines, and angles.

Gradient
The steepness of a line.

Graph
A diagram used to represent information, including the relationship between two sets of variables.

Greater than
An amount larger than another quantity. It is represented by the symbol >.

Greater than or equal to
An amount either larger or the same as another quantity. It is represented by the symbol ⩾.

Height
The upwards length, measuring between the lowest and highest points.

Hexagon
A two-dimensional shape with 6 sides.

Highest common factor
The largest number that divides exactly into a set of other numbers. It is often written as HCF, e.g. the HCF of 12 and 18 is 6.

Histogram
A graph that uses area to measure frequency.

Horizontal
Parallel to the horizon. A horizontal line goes between left and right.

Hypotenuse
The side opposite the right-angle in a right-angled triangle. It is the longest side of a right-angled triangle.

Impossibility
Something that could never happen. The probability of an impossibility is written as 0.

Improper fraction
Fraction in which the numerator is greater than the denominator.

Included angle
An angle formed between two sides with a common vertex.

Income
An amount of money earned.

Independent events
Occurrences that have no influence on each other.

Indices (singular: index)
See power.

Indirect proportion
Two variables x and y are in indirect proportion if e.g. when one variable doubles, the other halves, or vice versa.

Inequalities
Inequalities show that two statements are not equal.

Infinite
Without a limit or end. Infinity is represented by the symbol ∞.

Integers
Whole numbers that can be positive, negative, or zero, e.g. -3, -1, 0, 2, 6.

Interest
An amount of money charged when money is borrowed, or the amount earned when it is invested. It is usually written as a percentage.

Interior angle
1. An included angle in a polygon.
2. An angle formed when two lines are intersected by another line.

Intercept
The point on a graph at which a line crosses an axis.

Interquartile range
A measure of the spread of a set of data. It is the difference between the lower and upper quartiles.

Intersection/intersect
A point where two or more lines or figures meet.

Inverse
The opposite of something, e.g. division is the inverse of multiplication and vice versa.

Investment/invest
An amount of money spent in an attempt to make a profit.

Isosceles triangle
A triangle with two equal sides and two equal angles.

Kite
A quadrilateral made of two pairs of adjacent sides of equal length.

Length
The measurement of the distance between two points e.g. how long a line segment is between its two ends.

Less than
An amount smaller than another quantity. It is represented by the symbol <.

Less than or equal to
An amount smaller or the same as another quantity. It is represented by the symbol ⩽.

Like terms
An expression in algebra that contains the same symbols, such as x and y, (the numbers in front of the x or y may change). Like terms can be combined.

Line
A one-dimensional element that only has length (i.e. no width or height).

Line graph
A graph where points are connected by straight lines.

Line of best fit
A line on a scatter diagram that shows the correlation or trend between variables.

Line of symmetry
A line that acts like a mirror, splitting a figure into two mirror-image parts.

Loan
An amount of money borrowed that has to be paid back (usually over a period of time).

Locus (plural: loci)
The path of a point, following certain conditions or rules.

Loss
Spending more money than has been earned creates a loss.

Lowest common multiple
The smallest number that can be divided exactly into a set of values. It is often written LCM, e.g. the LCM of 4 and 6 is 12.

Major
The larger of the two or more objects referred to. It can be applied to arcs, segments, sectors, or ellipses.

Mean
The middle value of a set of data, found by adding up all the values, then dividing by the total number of values.

Measurement
A quantity, length, or size, found by measuring something.

Median
The number that lies in the middle of a set of data, after the data has been put into

increasing order. The median is a type of average.

Mental arithmetic
Basic calculations done without writing anything down.

Minor
The smaller of the two or more objects it referred to. It can be applied to arcs, segments, sectors, or ellipses.

Minus
The sign for subtraction, represented as –.

Mixed operations
A combination of different actions used in a calculation, such as addition, subtraction, multiplication, and division.

Mode
The number that appears most often in a set of data. The mode is a type of average.

Mortgage
An agreement to borrow money to pay for a house. It is paid back with interest over a long period of time.

Multiply/multiplication
The process of adding a value to itself a set number of times. The symbol for multiplication is ×.

Mutually exclusive events
Two mutally exclusive events are events that cannot both be true at the same time.

Negative
Less than zero. Negative is the opposite of positive.

Net
A flat shape that can be folded to make a three-dimensional object.

Not equal to
Not of the same value. Not equal to is represented by the symbol ≠, e.g. 1 ≠ 2.

Numerator
The number at the top of a

fraction, e.g. 2 is the numerator of $^2/_3$.

Obtuse angle
An angle measuring between 90° and 180°.

Octagon
A two-dimensional shape with 8 sides and 8 angles.

Odd number
A whole number that cannot be divided by 2, e.g. -7, 1, and 65.

Operation
An action done to a number, e.g. adding, subtracting, dividing, and multiplying.

Operator
A symbol that represents an operation, e.g. +, –, ×, and ÷.

Opposite
Angles or sides are opposite if they face each other.

Parallel
Two lines are parallel if they are always the same distance apart.

Parallelogram
A quadrilateral which has opposite sides that are equal and parallel to each other.

Pascal's triangle
A number pattern formed in a triangle. Each number is the sum of the two numbers directly above it. The number at the top is 1.

Pentagon
A two-dimensional shape that has 5 sides and 5 angles.

Percentage/per cent
A number of parts out of a hundred. Percentage is represented by the symbol %.

Perimeter
The boundary all the way around a shape. The perimeter also refers to the length of this boundary.

Perpendicular bisector
A line that cuts another line in half at right-angles to it.

Pi
A number that is approximately 3.142 and is represented by the Greek letter pi, π.

Pie chart
A circular graph in which segments represent different quantities.

Plane
A completely flat surface that can be horizontal, vertical, or sloping.

Plus
The sign for addition, represented as +.

Point of contact
The place where two or more lines intersect or touch.

Polygon
A two-dimensional shape with 3 or more straight sides.

Polyhedron
A three-dimensional object with faces that are flat polygons.

Positive
More than zero. Positive is the opposite of negative.

Power
The number that indicates how many times a number is multiplied by itself. Powers are shown by a small number at the top-right hand corner of another number, e.g. 4 is the power in 2^4 = 2 x 2 x 2 x 2.

Prime number
A number which has exactly two factors: 1 and itself. The first 10 prime numbers are 2, 3, 5, 7, 11, 13, 17, 19, 23, and 29.

Prism
A three-dimensional object with ends that are identical polygons.

Probability
The likelihood that something will happen. This likelihood is given a value between 0 and 1. An impossible event has probability 0 and a certain event has probability 1.

Product
A number calculated when two or more numbers are multiplied together.

Profit
The amount of money left once costs have been paid.

Proper fraction
A fraction in which the numerator is less than the denominator, e.g. $^2/_5$ is a proper fraction.

Proportion/proportionality
Proportionality is when two or more quantities are related by a constant ratio, e.g. a recipe may contain three parts of one ingredient to two parts of another.

Protractor
A tool used to measure angles.

Pyramid
A three-dimensional object with a polygon as its base and triangular sides that meet in a point at the top.

Pythagoras' theorem
A rule that states that the squared length of the hypotenuse of a right-angled triangle will equal the sum of the squares of the other two sides as represented by the equation $a^2 + b^2 = c^2$.

Quadrant
A quarter of a circle, or a quarter of a graph divided by the x- and y-axes.

Quadratic equation
Equations that include a squared variable, e.g. $x^2 + 3x + 2 = 0$.

Quadratic formula
A formula that allows any quadratic equation to be solved, by substituting values into it.

Quadrilateral
A two-dimensional shape that has 4 sides and 4 angles.

Quartiles
In statistics, quartiles are points that split an ordered set of data into 4 equal parts. The number that is a quarter of the way through is the lower quartile, halfway is the median, and three-quarters of the way through is the upper quartile.

Quotient
The whole number of times a number can be divided into another e.g. if $11 \div 2$ then the quotient is 5 (and the remainder is 1).

Radius (plural: radii)
The distance from the centre of a circle to any point on its circumference.

Random
Something that has no special pattern in it, but has happened by chance.

Range
The span between the smallest and largest values in a set of data.

Ratio
A comparison of two numbers, written either side of the symbol: e.g. 2:3.

Rectangle
A quadrilateral with 2 pairs of opposite, parallel sides that are equal in length, and 4 right angles.

Recurring
Something that repeats over and over again, e.g. $^1/_9 = 0.11111...$ is a recurring decimal and is shown as $0.\dot{1}$.

Reflection
A type of transformation that produces a mirror-image of the original object.

Reflex angle
An angle between 180° and 360°.

Regular polygon
A two-dimensional shape with sides that are all the same length and angles that are all the same size.

Remainder
The number left over when a dividing a number into whole parts e.g. $11 \div 2 = 5$ with remainder 1.

Revolution
A complete turn of 360°.

Rhombus
A quadrilateral with 2 pairs of parallel sides and all 4 sides of the same length.

Right angle
An angle measuring exactly 90°.

Root
The number which, when multiplied by itself a number of times, results in the given value, e.g. 2 is the fourth root of 16 as $2 \times 2 \times 2 \times 2 = 16$.

Rotation
A type of transformation in which an object is turned around a point.

Rounding
The process of approximating a number by writing it to the nearest whole number or to a given number of decimal places.

Salary
An amount of money paid regularly for the work that someone has done.

Sample
A part of a whole group from which data is collected to give information about the whole group.

Savings
An amount of money kept aside or invested and not spent.

Scale/scale drawing
Scale is the amount by which an object is made larger or smaller. It is represented as a ratio. A scale drawing is a drawing that is in direct proportion to the object it represents.

Scalene triangle
A triangle where every side is a different length and every angle is a different size.

Scatter diagram
A graph in which plotted points or dots are used to show the correlation or relationship between two sets of data.

Sector
Part of a circle, with edges that are two radii and an arc.

Segment
Part of a circle, whose edges are a chord and an arc.

Semi-circle
Half of a full circle, whose edges are the diameter and an arc.

Sequence
A list of numbers ordered according to a rule.

Similar
Shapes are similar if they have the same shape but not the same size.

Simplification
In algebra, writing something in its most basic or simple form, e.g. by cancelling terms.

Simultaneous equation
Two or more equations that must be solved at the same time.

Sine
In trigonometry, sine is the ratio of the side opposite to a given angle with the hypotenuse of a right-angled triangle.

Solid
A three-dimensional shape that has length, width, and height.

Sphere
A three-dimensional, ball-shaped, perfectly round object, where each point on its surface is the same distance from its centre.

Spread
The spread of a set of data is how the data is distributed over a range.

Square
A quadrilateral in which all the angles are the same (90°) and every side is the same length.

Square root
A number that, multiplied by itself, produces a given number, shown as $\sqrt{}$, e.g. $\sqrt{4} = 2$.

Squared number
The result of multiplying a number by itself, e.g. $4^2 = 4 \times 4 = 16$.

Standard deviation
A measure of spread that shows the amount of deviation from the mean. If the standard deviation is low the data is close to the mean, if it is high, it is widely spread.

Standard form
A number (usually very large or very small) written as a positive or negative number between 1 and 9 multiplied by a power of 10, e.g. 0.02 is 2×10^{-2}.

Statistics
The collection, presentation, and interpretation of data.

Stem-and-leaf diagram
A graph showing the shape of ordered data. Numbers are split in two digits and separated by a line. The first digits form the stem (written once) and the second digits form leaves (written many times in rows).

Substitution
Putting something in place of something else, e.g. using a constant number in place of a variable.

Subtraction/subtract
Taking a number away from another number. It is represented by the symbol –.

Sum
The total, or the number calculated when two numbers are added together.

Supplementary angle
Two angles that add up to 180°.

Symmetry/symmetrical
A shape or object is symmetrical if it looks the same after a reflection or a rotation.

Table
Information displayed in rows and columns.

Take-home pay
Take-home pay is the amount of earnings left after tax has been paid.

Tangent
1. A straight line that touches a curve at one point.
2. In trigonometry, tangent is the ratio of the side opposite to a given angle with the side adjacent to the given angle, in a right-angled triangle.

Tax
Money that is paid to the government, either as part of what a person buys, or as a part of their income.

Terms
Individual numbers in a sequence or series, or individual parts of an expression, e.g. in $7a^2 + 4xy - 5$ the terms are $7a^2$, $4xy$, and 5.

Tessellation
A pattern of shapes covering a surface without leaving any gaps.

Theoretical probability
The likelihood of an outcome based on mathematical ideas rather than experiments.

Three-dimensional
Objects that have length, width, and height. Three-dimensions is often written as 3D.

Transformation
A change of position, size, or orientation. Reflections, rotations, enlargements, and translations are all transformations.

Translation
Movement of an object without it being rotated.

Trapezium
A quadrilateral with a pair of parallel sides that can be of different lengths.

Triangle
A two-dimensional shape with 3 sides and 3 angles.

Trigonometry
The study of triangles and the ratios of their sides and angles.

Two-dimensional
A flat figure that has length and width. Two-dimensions is often written as 2D.

Unit
1. The standard amount in measuring, e.g. cm, kg, and seconds.
2. Another name for one and refers to the digit to the left of the decimal point.

Unknown angle
An angle which is not specified, and for which the number of degrees need to be determined.

Variable
A quantity that can vary or change and is usually indicated by a letter.

Vector
A quantity that has both size and direction, e.g. velocity and force are vectors.

Velocity
The speed and direction in which something is moving, measured in metres per second m/s.

Vertex (plural: vertices)
The corner or point at which surfaces or lines meet.

Vertical
At right-angles to the horizon. A vertical line goes between up and down directions.

Volume
The amount of space within a three-dimensional object. Volume is measured in units cubed, e.g. cm^3.

Wage
The amount of money paid to a person in exchange for work.

Whole number
Counting numbers that do not have any fractional parts and are greater than or equal to 0, e.g. 1, 7, 46, 108.

Whole turn
A rotation of 360°, so that an object faces the same direction it started from.

Width
The sideways length, measuring between opposite sides. Width is the same as breadth.

X-axis
The horizontal axis of a graph, which determines the x-coordinate.

X-intercept
The value at which a line crosses the x-axis on a graph.

Y-axis
The vertical axis of a graph, which determines the y-coordinate.

Index

A

accuracy 63
acute-angled triangles, area 115
acute angles 77, 241
addition 16
  calculators 64
  expressions 164
  fractions 45
  inequalities 190
  multiplication 18
  negative numbers 30
  positive numbers 30
  vectors 88
allowance, personal finance 66
algebra 158–91
alternate angles 79, 241
angle of rotation 92, 93, 241
angles 76–77
  45° 105
  60° 105
  90° 105
  acute 77
  alternate 79
  arcs 142
  bearings 100
  bisecting 104, 105
  complementary 77, 241
  congruent triangles 112, 113
  constructions 102
  corresponding 79
  cyclic quadrilaterals 139
  drawing triangles 110, 111
  in a circle 136–37
  obtuse 77, 241
  parallel lines 79, 241
  pie charts 202
  polygons 126, 127, 128
  protractor 74, 75
  quadrilaterals 122, 123
  reflex 77
  rhombus 125
  right-angled 77, 105, 241
  sectors 143
  supplementary 77, 241
  tangents 141
  triangles 108, 109
  trigonometry formulas 154, 155,
    156–57
annotation, pie charts 203
answer, calculator 65
approximately equals sign 62

approximation 62
arcs 130, 131, 142
  compasses 74
  sectors 143
area,
  circles 130, 131, 134–35, 143,
    147
  congruent triangles 112
  cross-sections 146
  formula 169
  measurement 28
  quadrilaterals 124–25
  rectangles 28
  triangles 114–16
arithmetic keys, calculators 64
arrowheads 78
averages 206–07
  frequency tables 208
  moving 210–11
axes,
  bar charts 198
  graphs 84, 176, 204, 205
axis of reflection 94, 95
axis of symmetry 81

B

balancing equations 172
banks, personal finance 66, 67
bar charts 198–201, 216
bearings 100–01
bias 197
bisectors 104, 105
  angles 104, 105
  perpendicular 102, 103, 138,
    139
  rotation 93
borrowing, personal finance 66,
  67
box-and-whisker diagrams 215
box method of multiplication 21
brackets,
  calculators 64, 65
  expanding expressions 166
break-even, finance 66, 68
business finance 68–69

C

calculators 64–65, 75
  cosine (cos) 154, 156
  exponent button 33

powers 33
  roots 33
  sine (sin) 154, 156
  standard form 37
  tangent (tan) 154, 156
calendars 28
cancel key, calculators 64
cancellation,
  equations 172
  expressions 165
  fractions 43, 56
  formulas 170
  ratios 48
capital 67
carrying numbers 24
Celsius temperature scale 177,
  239
centimetres 28, 29
centre of a circle 130, 131
  angles in a circle 136
  arcs 142
  chords 138, 139
  pie charts 203
  tangents 140, 141
centre of enlargement 96, 97
centre of rotation 81, 92, 93
chance 222, 223, 226, 228, 229
chances,
  dependent events 228, 229
  expectation 224
change,
  percentages 55
  proportion 50
charts 197
chords 130, 131, 138–39
  tangents 141
circles 130–31
  angles in a 76, 77, 136–37
  arcs 142
  area of 134–35, 143, 146, 147
  chords 130, 131, 138–39
  circumference 132
  compasses 74
  cyclic quadrilaterals 139
  diameter 132, 133
  formulas 241
  loci 106
  pie charts 202, 203
  sectors 143
  symmetry 80
  tangents 140, 141
circular prism 144

circumference 130, 131, 132
  angles in a circle 136, 137
  arcs 142
  chords 138
  cyclic quadrilaterals 139
  pie charts 203
  tangents 140, 141
codes 27
collinear points 78
common denominator 44–45
  ratio fractions 49
common factors 166, 167
common multiples 20
comparing ratios 48, 49
compass directions 100
compass points 100
compasses (for drawing circles)
  131
  constructing tangents 141
  constructions 102
  drawing a pie chart 203
  drawing triangles 110, 111
  geometry tools 74
complementary angles 77, 241
component bar charts 201
composite bar charts 201
composite numbers 26, 27
compound bar charts 201
compound interest 67
compound measurement units
  28
compound shapes 135
computer animation 110
concave polygons 128
cones 145
  surface area 149
  volumes 147
congruent triangles 104, 112–13
  drawing 110
  parallelograms 125
constructing reflections 95
constructing tangents 141
constructions 102–03
convex polygon 128, 129
coordinates 82–83
  constructing reflections 95
  enlargements 97
  equations 85, 180, 181, 187, 189
  graphs 84, 174
  linear graphs 174
  maps 85
  quadratic equations 187, 189

rotation 93
simultaneous equations 180, 181
correlations, scatter diagrams 218, 219
corresponding angles 79
cosine (cos),
 calculators 65
 formula 154, 155, 156, 157
costs 66, 68, 69
credit 66
cross-sections,
 solids 144
 volumes 146
cube roots 33
 estimating 35
cubed numbers,
 calculator 65
 powers 32
 units 28
cubic units 146
cuboids 144, 145
 surface area 149
 symmetry 80, 81
 volume 28, 147
cubes 145
cumulative frequency graphs 205
 quartiles 214
curves, quadratic equation
 graphs 186
cyclic quadrilaterals 138, 139
cylinders 144, 145
 nets 148
 surface area 148, 167
 symmetry 81
 volume 146

D

data 196–97
 averages 206, 207, 210–11
 bar charts 198, 199, 200, 201
 cumulative frequency graphs 205
 frequency tables 208
 grouped 209
 leaf diagrams 213
 line graphs 204
 moving averages 210–11
 quartiles 214, 215
 ratios 48
 scatter diagrams 218, 219

spread 212
stem diagrams 213
data logging 197
data presentation,
 histograms 216, 217
 pie charts 202
data protection 27
data table 200
days 28
decagons 127
decimal numbers 38–39
 converting 56–57
 division 24, 25
 mental mathematics 59
decimal places,
 rounding off 63
 standard form 36
decimal points 38
 calculators 64
 standard form 36
decrease as percentages 55
degrees,
 angles 76
 bearings 100
deletion, calculators 64
denominators,
 adding fractions 45
 common 44–45
 fractions 40, 41, 42, 43, 45, 56, 57
 ratio fractions 49
 subtracting fractions 45
density measurement 28, 29
dependent events 228–29
 tree diagrams 231
diagonals in quadrilaterals 122, 123
diameter 130, 131, 132, 133
 angles in a circle 137
 area of a circle 134, 135
 chords 138
difference, subtraction 17
direct proportion 50
direction,
 bearings 100
 vectors 86
distance,
 bearings 101
 loci 106
 measurement 28, 29
distribution,
 data 212, 231

quartiles 214, 215
dividend 22, 23, 24, 25
division 22–23
 calculators 64
 cancellation 43
 decimal numbers 39
 expressions 165
 formulas 170
 fractions 42, 47
 inequalities 190
 long 25
 negative numbers 31
 positive numbers 31
 powers 34
 proportional quantities 51
 quick methods 60
 ratios 49, 51
 short 24
 top-heavy fractions 42
divisor 22, 23, 24, 25
dodecagons 126, 127
double inequalities 191
double negatives 65
drawing constructions 102
drawing triangles 110–11

E

earnings 66
edges of solids 145
eighth fraction 40
elimination, simultaneous
 equations 178
employees, finance 68
employment, finance 66
encryption 27
endpoints 78
enlargements 96–97
equal vectors 87
equals sign 16, 17
 approximately 62
 calculators 64
 equations 172
 formulas 169
equations,
 coordinates 85
 factorizing quadratic 182–83
 graphs 186, 187
 linear graphs 174, 175, 176, 177
 Pythagoras' theorem 120, 121
 quadratic 182–85, 186, 187
 simultaneous 178–81

solving 172–73
equiangular polygons 126
equilateral polygons 126
equilateral triangle 105, 109
 symmetry 80, 81
equivalent fractions 43
estimating,
 calculators 64
 cube roots 35
 quartiles 214
 rounding off 62
 square roots 35
Euclid 26
evaluating expressions 165
even chance 223
expanding expressions 166
expectation 224–25
exponent button, calculators 33, 37, 65
expressions 164–65
 equations 172
 expanding 166–67
 factorizing 166–67
 quadratic 168
 sequences 162
exterior angles,
 cyclic quadrilaterals 139
 polygons 129
 triangles 109

F

faces of solids 145, 148
factorizing 27
 expressions 166, 167, 168
 quadratic equations 182–83
 quadratic expressions 168
factors 166, 167
 division 24
 prime 26, 27
Fahrenheit temperature scale 177
feet 28
Fibonacci sequence 163, 239
finance,
 business 68–67
 personal 66–67
flat shapes, symmetry 80
formulas 169
 area of quadrilaterals 124
 area of rectangles 165
 area of triangles 114, 115, 116

factorizing 166
interest 67
moving terms 170–71
Pythagoras' theorem 120, 121
quadratic equations 183,
  184–85
quartiles 214, 215
speed 29
trigonometry 154
fractional numbers 38
fractions 40–47
  adding 45
  common denominators 44
  converting 56–57
  division 47
  mixed 42
  multiplication 46
  probability 222, 225, 226
  ratios 49
  subtracting 45
  top-heavy 42
frequency,
  bar charts 198, 199, 200
  cumulative 205
frequency density 216, 217
frequency graph 214
frequency polygons 201
frequency tables 208, 209
  bar charts 198, 199
  data presentation 197
  histograms 217
  pie charts 202
function keys, scientific
  calculator 65
functions, calculators 64, 65

G

geometry 70–149
geometry tools 74–75
government, personal finance 66
gradients, linear graphs 174, 175
grams 28, 29
graphs,
  coordinates 82, 84
  cumulative frequency 205
  data 197
  line 204–05
  linear 174–77
  moving averages 210–11
  proportion 50
  quadratic equations 186–89

quartiles 214
scatter diagrams 218, 219
simultaneous equations 178,
  180–81
greater than symbol 190
grouped data 209

H

half fraction 41
hendecagons 127
heptagons 127, 128
hexagons 126, 127, 129
  tessellations 91
histograms 216–17
horizontal bar chart 200
horizontal coordinates 82, 83
hours 28, 29
  kilometres per 29
hundreds,
  addition 16
  decimal numbers 38
  multiplication 21
  subtraction 17
hypotenuse 109
  congruent triangles 113
  Pythagoras' theorem 120, 121
  tangents 140
  trigonometry formulas 154, 155,
    156, 157

I

icosagons 127
imperial measurements 28
inches 28
included angle, congruent
  triangles 113
income 66
income tax 66
increase, percentages 55
independent events 228
inequalities 190–91
infinite symmetry 80
inputs, finance 68
interest 67
  formulas 171, 240
  personal finance 66
interior angles,
  cyclic quadrilaterals 139
  polygons 128, 129
  triangles 109

interquartile range 215
intersecting chords 138
intersecting lines 78
inverse cosine 156
inverse multiplication 22
inverse proportion 50
inverse sine 156
inverse tangent 156
investment 66
  interest 67
irregular polygons 126, 127, 128,
  129
irregular quadrilaterals 122
isosceles triangles 109, 113
  rhombus 125
  symmetry 80

K

kaleidoscopes 94
keys,
  calculators 64
  pie charts 203
kilograms 28
kilometres 28
kilometres per hour 29
kite quadrilaterals 122, 123

L

labels on pie charts 203
latitude 85
leaf diagrams 213
length measurement 28
  speed 29
less than symbol 190
like terms in expressions 164
line graphs 204–05
line of best fit 219
line of symmetry 95
line segments 78
  constructions 103
  vectors 86
linear equations 174, 175, 176,
  177
linear graphs 174–77
lines 78
  angles 76, 77
  constructions 102, 103
  loci 106
  of symmetry 80
  rulers 74, 75

straight 77, 78–79
loans 66
location 106
locus (loci) 106–07
long division 25
long multiplication 21
  decimal numbers 38
longitude 85
loss,
  business finance 68
  personal finance 66
lowest common denominator 44
lowest common multiple 20

M

magnitude, vectors 86, 87
major arcs 142
major sectors 143
map coordinates 82, 83, 85
mass measurement 28
  density 29
mean,
  averages 206, 207, 210, 211
  frequency tables 208
  grouped data 209
  moving averages 210, 211
  weighted 209
measurement,
  drawing triangles 110
  scale drawing 98, 99
  units of 28–29
measuring spread 212–13
median,
  averages 206, 207
  quartiles 214, 215
memory, calculators 64
mental mathematics 58–61
metres 28
metric measurement 28
miles 28
milliseconds 28
minor arcs 142
minor sectors 143
minus sign 30
  calculator 65
minutes 28, 29
mirror image,
  reflections 94
  symmetry 80
mixed fractions 41, 42, 46
  division 47

multiplication 46
modal class 209
mode 206
money 68
  business finance 69
  interest 67
  personal finance 66
months 28
mortgage 66
multiple bar chart 201
multiple choice questions 196
multiple probabilities 226–27
multiples 20
  division 24
multiplication 18–21
  calculators 64
  decimal numbers 38
  expanding expressions 166
  expressions 165
  formulas 170
  fractions 42, 46
  indirect proportion 50
  inequalities 190
  long 21
  mental mathematics 58
  mixed fractions 42
  negative numbers 31
  positive numbers 31
  powers 32, 34
  proportional quantities 51
  reverse cancellation 43
  short 21
  vectors 88

N

negative correlations 219
negative gradients 175
negative numbers 30–31
  addition 30
  calculators 65
  dividing 31
  inequalities 190
  multiplying 31
  quadratic graphs 187
  subtraction 30
negative scale factor 96
negative terms in formulas 170
negative translation 91
negative values on graphs 84
negative vectors 87
nets 144, 148, 149

non parallel lines 78
non-polyhedrons 145
nonagons 127, 129
nought 30
"nth" value 162
number line,
  addition 16
  negative numbers 30-31
  positive numbers 30-31
  subtraction 17
numbers,
  calculators 64
  composite 26
  decimal 38–39
  negative 30–31
  positive 30–31
  prime 26–27
numerator 40, 41, 42, 43, 56, 57
  adding fractions 45
  comparing fractions 44
  ratio fractions 49
  subtracting fractions 45

O

obtuse-angled triangle 115
obtuse angles 77
obtuse triangle 109
octagons 127
operations,
  calculators 65
  expressions 164
order of rotational symmetry 81
origin 84
ounces 28
outputs, business finance 68, 69
overdraft 66

P

parallel lines 78, 79
  angles 79
parallel sides of a parallelogram
  125
parallelograms 78, 122, 123
  area 125
patterns,
  sequences 162
  tessellations 91
pension plan 66
pentadecagon 127
pentagons 127, 128, 129

symmetry 80
pentagonal prism 144
percentages 52–55
  converting 56–57
  interest 67
  mental mathematics 61
perimeters,
  circles 131
  triangles 108
perpendicular bisectors 102, 103
  chords 138, 139
  rotation 93
perpendicular (vertical) height,
  area of quadrilaterals 124, 125
  area of triangles 114, 115
  volumes 147
perpendicular lines,
  constructions 102, 103
personal finance 66–67
personal identification number
  (PIN) 66
pi (π) 132, 133
  surface area of a cylinder 167
  surface area of a sphere 149
  volume of sphere 147
pie charts 202–03
  business finance 69
planes 78
  symmetry 80
  tessellations 91
plotting,
  bearings 100, 101
  enlargements 97
  graphs 84
  line graphs 204
  linear graphs 176
  loci 107
  simultaneous equations 180,
    181
plus sign 30
points,
  angles 76, 77
  constructions 102, 103
  lines 78
  loci 106
  polygons 126
polygons 126–29
  enlargements 96, 97
  frequency 201
  irregular 126, 127
  quadrilaterals 122
  regular 126, 127

triangles 108
polyhedrons 144, 145
positive correlation 218, 219
positive gradients 175
positive numbers 30–31
  addition 30
  dividing 31
  inequalities 190
  multiplying 31
  quadratic graphs 187
  subtraction 30
positive scale factor 96
positive terms in formulas 170
positive translation 91
positive values on graphs 84
positive vectors 87
pounds (mass) 28
power of ten 36, 37
power of zero 34
powers 32
  calculators 65
  dividing 34
  multiplying 34
prime factors 26, 27
prime numbers 26–27
prisms 144, 145
probabilities, multiple
  226–27
probability 222–23
  dependent events 228
  expectation 224, 225
  tree diagrams 230
probability fraction 225
probability scale 222
processing costs 69
product,
  business finance 68
  indirect proportion 50
  multiples 20
  multiplication 18
profit,
  business finance 68, 69
  personal finance 66
progression, mental
  mathematics 61
proper fractions 41
  division 47
  multiplication 46
properties of triangles 109
proportion 48, 50
  arcs 142
  enlargements 96

percentages 54, 56
sectors 143
similar triangles 117, 119
proportional quantities 51
protractors,
  drawing pie charts 203
  drawing triangles 110, 111
  geometry tools 74, 75
  measuring bearings 100,
    101
pyramids 145
  symmetry 80, 81
Pythagoras' theorem 120–21
  tangents 140
  vectors 87

Q

quadrants, graphs 84
quadratic equations 184–85
  factorizing 182–83
  graphs 186–89
quadratic expressions 168
quadratic formulas 184–85
quadrilaterals 122–25, 128
  area 124–25
  cyclic 138, 139
  polygons 127
quantities,
  proportion 48, 50, 51
  ratio 48
quarter fraction 40
quartiles 214–15
quotient 22, 23
  division 25

R

radius (radii) 130, 131, 132, 133
  area of a circle 134, 135
  compasses 74
  sectors 143
  tangents 140
  volumes 147
range,
  data 212, 213
  histograms 217
  quartiles 214
rate, interest 67
ratio 48–49, 50
  arcs 142
  scale drawing 98

similar triangles 118, 119
  triangles 51, 118, 119
raw data 196
reality 224–25
recall button, calculators 64
rectangles,
  area of 28, 124, 165
  polygons 126
  quadrilaterals 122, 123
  symmetry 80
rectangle-based pyramid 80, 81
rectangular prism 144
recurring decimal numbers 39
re-entrant polygons 126
reflections 94–95
  congruent triangles 112
reflective symmetry 80
  circles 130
reflex angles 77
  polygons 128
regular pentagons 80
regular polygons 126, 127, 128,
  129
regular quadrilaterals 122
relationships, proportion 50
remainders 23, 24, 25
revenue 66, 68, 69
reverse cancellation 43
rhombus,
  angles 125
  area of 124
  polygons 126
  quadrilaterals 122, 123, 124
right angles 77
  angles in a circle 137
  congruent triangles 113
  constructing 105
  hypotenuse 113
  perpendicular lines 102
  quadrilaterals 122, 123
right-angled triangles 109
  calculators 65
  Pythagoras' theorem 120, 121
  set squares 75
  tangents 140
  trigonometry formulas 154, 155,
    156, 157
  vectors 87
roots 32, 33
rotational symmetry 80, 81
  circles 130
rotations 92–93

congruent triangles 112
rounding off 62–63
rulers,
  drawing a pie chart 203
  drawing circles 131
  drawing triangles 110, 111
  geometry tools 74, 75

S

sales tax 66
savings, personal finance
  66, 67
scale,
  bar charts 198
  bearings 101
  drawing 98–99
  probability 222
  ratios 49
scale drawing 98–99
scale factor 96, 97
scalene triangle 109
scaling down 49, 98
scaling up 49, 98
scatter diagrams 218–19
scientific calculators 65
seasonality 210
seconds 28
sectors 130, 131, 143
segments,
  circles 130, 131
  pie charts 202
seismometer 197
sequences 162–63
series 162
set squares 75
shapes,
  compound 135
  constructions 102
  loci 106
  polygons 126
  quadrilaterals 122
  solids 144
  symmetry 80, 81
  tessellations 91
shares 66
sharing 22
short division 24
short multiplication 21
sides,
  congruent triangles 112, 113
  drawing triangles 110, 111

polygons 126, 127
quadrilaterals 122, 123
triangles 108, 109, 110, 111, 112,
  113, 154–55, 156, 157
signs (see also symbols),
  addition 16
  approximately equals 62
  equals 16, 17, 64, 169
  minus 30, 65
  multiplication 18
  negative numbers 30, 31
  plus 30
  positive numbers 30, 31
  subtraction 17
significant figures 63
similar triangles 117–19
simple equations 172
simple interest 67
  formula 171
simplifying,
  equations 172, 173
  expressions 164–65
simultaneous equations 178–81
sine,
  calculators 65
  formula 154, 156
size,
  measurement 28
  ratio 48
  vectors 86
solids 144–45
  surface areas 144, 148–49
  symmetry 80
  volumes 146
solving equations 172–73
solving inequalities 191
speed measurement 28, 29
spheres,
  solids 145
  surface area 149
  volume 147
spirals,
  Fibonacci sequence 163
  loci 107
spread 212–13
  quartiles 215
square numbers sequence 163
square roots 33
  calculators 65
  estimating 35
  Pythagoras' theorem 121
square units 28, 124

squared numbers,
 powers 32
 quadratic equations 184
squared variables,
 quadratic equations 182
 quadratic expressions 168
squares,
 area of quadrilaterals 124
 calculators 65
 polygons 126
 quadrilaterals 122, 123
 symmetry 80, 81
 tessellations 91
squaring,
 expanding expressions 166
 Pythagoras' theorem 120
standard form 36–37
statistics 192–219
stem diagrams 213
straight lines 78–79
 angles 77
subject of a formula 169
substitution,
 equations 172, 178, 179, 184
 expressions 165
 quadratic equations 184
 simultaneous equations 178,
  179
subtended angles 136, 137
subtraction 17
 calculators 64
 expressions 164
 fractions 45
 inequalities 190
 negative numbers 30
 positive numbers 30
 vectors 88
sums 16
 calculators 64, 65
 multiplication 18, 19
supplementary angles 77
surface area,
 cylinder 167
 solids 144, 148–49
surveys, data collection 196–97
switch, mental mathematics 61
symbols (*see also* signs),
 cube roots 33
 division 22
 expressions 164, 165
 greater than 190
 inequality 190

less than 190
 ratio 48, 98
 square roots 33
 triangles 108
symmetry 80–81
 circles 130

T

table of data 218
 pie charts 202
tables,
 data 200
 data collection 196, 197
 frequency 198, 199, 208
 proportion 50
taking away (subtraction) 17
tally charts 197
tangent formula 154, 155, 156,
 157
tangents 130, 131, 140–41
 calculators 65
tax 66
temperature 31
temperature conversion graph
 177
tens,
 addition 16
 decimal numbers 38
 multiplication 21
 subtraction 17
tenths 38
terms,
 expressions 164, 165
 moving 170
 sequences 162
tessellations 91
thermometers 31
thousands,
 addition 16
 decimal numbers 38
three-dimensional bar chart
 200
three-dimensional shapes
 144
 symmetry 80, 81
time measurement 28
 speed 29
times tables 59
tonnes 28
top-heavy fractions 41, 42, 46
transformations,

enlargements 96
 reflections 94
 rotation 92
 translation 90
translation 90–91
transversals 78, 79
trapezium (trapezoid) 122, 123,
 126
tree diagrams 230–31
triangles 108–09
 area of 114–16
 calculators 65
 congruent 104, 125
 constructing 110–11
 equilateral 105
 formulas 29, 169
 rhombus 125
 parallelograms 125
 polygons 126, 127
 Pythagoras' theorem 120,
  121
 right-angled 65, 75, 87, 109, 120,
  121, 140, 154, 155, 156, 157
 set squares 75
 similar 117–19
 symmetry 80, 81
 tangents 140
 trigonometry formulas 154, 155,
  156, 157
 vectors 87, 89
trigonometry, 150-57
 calculators 65
 formulas 154–57
turns, angles 76
two-dimensional shapes,
 symmetry 80, 81
two-way table 197

U

units (numbers),
 addition 16
 decimal numbers 38
 multiplication 21
 subtraction 17
units of measurement
 28–29
 cubed 146
 ratios 49
 squared 124
unsolvable simultaneous
 equations 181

V

variables,
 equations 172
 simultaneous equations 178,
  179
vectors 86–89
 translation 90, 91
vertex (vertices) 108
 angles 77
 bisecting an angle 104
 cyclic quadrilaterals 139
 polygons 126
 quadrilaterals 122, 139
 solids 145
vertical coordinates 82, 83
vertical (perpendicular) height,
 area of quadrilaterals 124, 125
 area of triangles 114, 115
 volumes 147
vertically-opposite angles 79
volume 144, 146–47
 density 29
 measurement 28

W

wages 66
weight measurement 28
weighted mean 209

X

x axis,
 bar charts 198, 199
 graphs 84

Y

y axis,
 bar charts 198, 199
 graphs 84
yards 28
years 28

Z

zero 30
zero correlations 219
zero power 34

Acknowledgements

BARRY LEWIS would like to thank Toby, Lara, and Emily, for always asking why.

DORLING KINDERSLEY would like to thank: David Summers, Cressida Tuson, and Ruth O'Rourke-Jones for additional editorial work, Kenny Grant, Sunita Gahir, Peter Laws, Steve Woosnam-Savage, and Hugh Schermuly for additional design work. We would also like to thank Sarah Broadbent for her work on the glossary.

For more information about Carol Vorderman's online maths school see **www.themathsfactor.com**

themathsfactor.com
with **Carol Vorderman**

The publisher would like to thank the following for their kind permission to reproduce their photographs:

(Key: b-bottom; c-centre; l-left; r-right; t-top)

Alamy Images: Bon Appetit 210bc (tub); K-PHOTOS 210bc (cone); **Corbis:** Doug Landreth/Science Faction 163cr; Charles O'Rear 197br; **Dorling Kindersley:** NASA 37tr, 85bl, 223br; Lindsey Stock 27br, 212cr; **Character from Halo 2 used with permission from Microsoft:** 110tr; **NASA:** JPL 37cr

All other images © Dorling Kindersley
For further information see: www.dkimages.com